Feedback

W0037274

Péter Érdi

Feedback

How to Destroy or Save the World

 Springer

Péter Érdi
Center for Complex Systems Studies
Kalamazoo College
Kalamazoo, MI, USA

ISBN 978-3-031-62438-4   ISBN 978-3-031-62439-1   (eBook)
https://doi.org/10.1007/978-3-031-62439-1

© The Editor(s) (if applicable) and The Author(s), under exclusive license to Springer Nature
Switzerland AG 2024

This work is subject to copyright. All rights are solely and exclusively licensed by the Publisher, whether
the whole or part of the material is concerned, specifically the rights of reprinting, reuse of illustrations,
recitation, broadcasting, reproduction on microfilms or in any other physical way, and transmission or
information storage and retrieval, electronic adaptation, computer software, or by similar or dissimilar
methodology now known or hereafter developed.
The use of general descriptive names, registered names, trademarks, service marks, etc. in this publication
does not imply, even in the absence of a specific statement, that such names are exempt from the relevant
protective laws and regulations and therefore free for general use.
The publisher, the authors and the editors are safe to assume that the advice and information in this book
are believed to be true and accurate at the date of publication. Neither the publisher nor the authors or
the editors give a warranty, expressed or implied, with respect to the material contained herein or for any
errors or omissions that may have been made. The publisher remains neutral with regard to jurisdictional
claims in published maps and institutional affiliations.

This Springer imprint is published by the registered company Springer Nature Switzerland AG
The registered company address is: Gewerbestrasse 11, 6330 Cham, Switzerland

If disposing of this product, please recycle the paper.

*I thank my former students **Brad Flaugher, Elliot Paquette, Griffin Drutchas, Jerrod Howlett,** and **Trevor Jones** for establishing the Interdisciplinary Fund for Complex Systems Studies at Kalamazoo College.*

Foreword

To lead into Péter Érdi's thought-provoking and highly readable *Feedback: How to Destroy or Save the World*, I discuss three feedback systems: the thermostat, the flush toilet—both parts of many readers' lives—and the helmsman.

A thermostat measures the error between a desired temperature and the actual temperature of a room. If the temperature is too low, the thermostat negates the error by a signal to turn on the heating system until the desired temperature is reached. A thermostat connected to both heating and air-conditioning can help negate the error whether the temperature is too high or too low. The thermostat is the classic example of a negative feedback system—it serves to detect some error and provide the control signal to negate, or at least reduce, it.

Now consider a couple who share a double bed and an electric blanket that has separate heating elements for each half with its own controller. Imagine if, by some happenstance, the controls get mixed. Here, when each person thinks they are controlling their half of the blanket they are instead controlling the other. When one person is feeling cold, they use the control on their side of the bed to turn up the heat—but, unfortunately, it is the heat on the other side. Their companion gets too hot and reaches for their control—only to making the bed even colder for the first person, who responds by turning up the heat on the other's side even more. Here we see an unfortunate case of what is called positive feedback, even though it has a negative effect. Rather than negating the error, it serves to increase it. This example can be extended to interpersonal relationships. When someone criticizes you, you may respond with soothing words, possibly calming the situation. In other cases, you get angry, setting off an expanding spiral of bad temper. We see here how feedback can work on a personal scale to destroy or save the relationship.

However, this book demonstrates that the mantra "negative feedback positive, positive feedback negative" is simplistic. At times, we need to encourage growth, and at other times, control it. More to the point, if we wish to avoid destroying the world, we cannot focus on feedback one system at a time. We must instead look at a multitude of interacting systems, where saving key systems may require multiple changes elsewhere.

To move from the interpersonal toward the global, let's start with the flush toilet! Again, we have a basic feedback mechanism which, detecting the low level of water in the cistern after each flush, signals adding water to refill the cistern to a preset level. But to be truly effective, the toilet needed a side invention—the S-shape of the exit pipe from the toilet bowl which ensures that after each flush no smelly material can float up into the bowl. We see here an example of embedding a negative feedback system within a larger system to get its full benefits. However, at their most effective, flush toilets are connected both to a reliable water supply system and a reliable sewage system. Such arrangements helped transform cities from places of fetid living and rampant disease to places that can support healthy lives. Of course, water and sewage are just two of the many interacting systems needed for a large city to support the well-being of its inhabitants, and not just the wealthy. Moreover, the people in the city—and the surrounding countryside and beyond—are crucial subsystems within the overall complexity. We see the increasing importance of feedback in ever-more complex systems as we move up the scale from humans to towns to cities to the whole planet—a level that has become increasingly a focus of the consciousness of many humans at this time of global warming.

Finally, the helmsman. Norbert Wiener had this figure in mind when he chose the title Cybernetics or Control and Communication in the Animal and the Machine for his amazingly influential book that established Cyber- as an almost ubiquitous prefix for the vocabulary of our twenty-first-century information age.

The term cybernetics comes from the Greek word kybernētēs: (helmsman, governor, or pilot). It is cognate with the Latin term gubernator for governor, and those who knew some Latin were amused when Arnold Schwarzenegger became governor of California by the similarity between gubernator and Terminator.

Enough frivolity! The helmsman controls the rudder of a ship as it navigates its way down the river or across the seas. He embodies the principle of negative feedback—if he sees the ship departing from its chosen path, he turns the rudder to move the ship back to its desired direction. However, the larger the ship the slower it will be to respond to the helmsman's feedback corrections. This can lead to many types of system malfunction unless the helmsman plans ahead—whether a collision with another ship or the riverbank, or going into wildly swinging oscillations as the delayed effect of the correction signal yields an over-correction which must then be counteracted to yield an over-over-correction, and so it goes… badly. Thus, if we are to even begin to hope to save the world, we must not only judiciously balance positive and negative feedback in a vast network of interlocking systems, but our interventions must be timely. Will these interventions succeed? Alas, as Wiener knew full well, it is very hard to make predictions, especially about the future.

And so I end where Péter Èrdi begins—with the dreams of Norbert Wiener.

La Jolla, California, USA   Michael Arbib
March 2024

Michael Arbib used the word feedback in his first book, *Brains, Machines and Mathematics.* His recent books include *How the Brain Got Language* and *When Brains Meet Buildings: A Conversation Between Neuroscience and Architecture.*

Preface

This book, of course, is about feedback. The concept of feedback is ancient. However, similarly to Moliére's Monsieur Jourdain, who did not know that he had been speaking prose all his life, people who built systems capable of self-regulation did not even give names for their techniques. We call it feedback.

I had two motivations to write this book. I will start with the second one, which is much newer than the first. I realized that many of us are wrestling with the problem of the world's future. (We have already discussed some aspects of the problem in our previous book written with Zsuzsa Szvetelszky: Repair: When and How to Improve Broken Objects, Ourselves, and Our Society. "Many people now agree that something went wrong in the world. Food waste and hunger, cheap clothing for the rich manufactured in conditions close to slavery in another part of the world, climate crisis, and social inequality. More frequent natural and social disasters" [1].

Will humanity survive, and will our grandchildren (including Hanna and Leo) live in prosperity? Or should we worry about the possibility of the extinction of humanity? The book is based on one assumption, hypothesis, on a belief: There is a *narrow border* between destruction and prosperity: to ensure reasonable growth but avoid existential risk, we need to find a fine-tuned balance between positive and negative feedback. My attempt is not to deal with the impossible task, to prove (certainly not in the spirit of formal feedback control theory), but to support the belief.

The second (my initial) motivation is personal: my love for the somewhat ill-fated cybernetics. As it happens in love, we cannot always rationalize our feelings. Norbert Wiener (1894–1964) created the modern field of cybernetics. His book *Cybernetics: Control and Communication in the Animal and the Machine* was a pluralistic theory and an interdisciplinary movement of several leading intellectuals. The term cybernetics goes back to Plato, who explained the principles of political self-governance. (Wiener referred to the ancient Greeks but did not mention Plato by name). Feedback is the fundamental concept of cybernetics. It is a process whereby some proportion of the output signal of a system is passed (fed back) to the input. So, the system itself contains a loop. Feedback mechanisms fundamentally influence the dynamic behavior of a system. Roughly speaking, negative feedback reduces the deviation or error from a goal state and has stabilizing effects. Positive feedback, which increases the deviation from an initial state, has destabilizing effects. Natural, technological, and social systems are full of feedback mechanisms.

Cyberneticians realized that goal-oriented systems (designed machines and animals) don't operate on the "single cause, single effect" paradigm, but what is called *circular causality*. In essence, it is a sequence of causes and effects whereby the explanation of a pattern leads back to the first cause, and either confirms or changes that first cause. Example: A causes B causes C that causes or modifies A. The concept had a somewhat bad reputation in legitimate scientific circles since it was somehow related to vicious circles in reasoning. It was reintroduced to science by cybernetics emphasizing feedback. In a feedback system, there is no clear discrimination between causes and effects since the output influences the input. The book secretly celebrates the revival of the perspective of cybernetics. I am honored that Michael Arbib, one of the last students of the two Founding Fathers of Cybernetics (Wiener and Warren McCulloch [1898–1969]), accepted writing the Foreword.

The flow of the book. Chapter 1, *Norbert Wiener's Dream: Technology, Life, and Society* starts with his perspective to describe goal-oriented behaviors in machines and animals. Wiener had a deep interest in the future of human society. Initially, no sufficient tools (data, models, computers) existed to make predictions. General System Theory and the development of simulation tools led to the emergence of the celebrated "world models," leading to the report of Limits to Growth [2].

If the Reader accepts that (i) it is worth studying the "narrow border" hypothesis and (ii) the feedback control approach might be a beneficial strategy to avoid catastrophes, then a massive but non-technical summary of the topic "feedback everywhere" in Chaps. 2, 3, 4, and 5 will encourage the Reader to think with me about possible natural and social disasters and strategies to avoid them.

Chapter 2, *Feedback Control: A Modern Concept is (a not so) Invisible Thread in the History of Technology* was strongly motivated by the writings of Dennis Bernstein, a professor of Aerospace Engineering at the University of Michigan. He showed the power of the concept by reviewing studies on mechanical clocks, steam engines, aerodynamics, and electronic devices.

Chapter 3, *Feedback Control in Biology: An Overview* illustrates that the concept is a fundamental tool at every level of the biological hierarchy, from cellular to socio-ecological systems. Feedback control strategies maintain the stable, healthy operation of biological systems. Stability has nothing to do with time-independent, stationary behavior, and stable, self-sustained oscillations with very different parameters have indispensable physiological roles.

Chapter 4, *Climate Changes, Wildfires, Tsunamis* analyzes the appearance of complicated feedback loops in generations of seemingly natural disasters. Climate predictions have their problems, and because of the presumably chaotic nature of the climate, predictability has severe limitations. At the global level, the future will tell whether social control will be sufficient to stop negative tendencies.

Chapter 5, *From Laissez-Faire to Greenspan: Feedback Control in Economic Systems* summarizes the central possible answers to the age-old question: Is economics a self-regulatory system that works reasonably, leaving everything to the decision of the individual participants (buyers and sellers), or should the government intervene. Symbolically, it can be labeled as the Keynes versus Friedman debate. Former Fed Chairman Greenspan realized some regulation was missing in the financial system. So far, humanity has not found any better suggestion that democratic institutions should have the power of control. Recent concerns emerged that the financial power of the few can be converted into political and legal power. We leave it for the next generation to stop and reverse this negative spiral.

As we know now how feedback control works in technology, biology, economics, and the global environment in Chap. 6 *From Natural Disasters to Social Riots*, we are ready to discuss the predictability of catastrophic and existential risks and the control mechanisms avoiding them. The complex systems approach to political instabilities combines causal modeling techniques and machine learning methods-based data processing to study the sophisticated problems: How much social unrest do we need? How much migration do we need? Even if we don't have direct answers, we should consider the stabilizing and destabilizing mechanisms to keep the world safe. Chapter 7 *Epilogue: The Narrow Border between Prosperity and Destruction* suggests that first, we need systems for providing early warnings for possible disasters. Second, rapid and local interventions by making decisions and taking actions may (or may not) save the world.

There are several excellent books discussing different aspects of feedback. A very tentative list [3–7]:

- Wiener N: The Human Use of Human Beings. Houghton Mifflin, 1950.
- Arthur, W. B. (1994). Increasing returns and path dependence in the economy. Ann Arbor, MI: University of Michigan Press.
- Diamond J: Collapse: How Societies Choose to Fail or Succeed. Viking Press 2004/2011.
- Tetlock PE and Gardner D: Superforecasting: The Art and Science of Prediction. Crown 2016.
- Cavana RY et al.: Feedback Economics: Economic Modeling with System Dynamics. Springer 2021.

(Last minute note: As I complete the Preface, I learned about the recent publication of a book *Feedback: Uncovering the Hidden Connections Between Life and the Universe* by Nicholas Golledge [8]. In that book, cycles of changes driven by feedback mechanisms are suggested to lead to life and creativity. I am sure the two books will reinforce each other.)

The book is intended to be an academic trade book. While the primary readership is mostly the broadly defined academics, it is not a research monograph. I hope some people outside Academia, say from public policy, might find some message. I see the book in the hands of people from various generations. Who are they? Young people are growing up in a world where everything seems to fall apart. People in their 30s who are thinking about how to live a fulfilling life. As recently written in a Forbes article, "Contrary to popular belief, Millennials read more than older generations do; and more than the last generation did at the same age." The text will also interest people in their 50s and 60s who are thinking back on life and how to repair their relationships. In 2022, about 70 million baby boomers in the United States will still be reflecting on their past successes and failures. Male Boomers, many of whom have a forever-young mentality, like to read non-fiction, which connects them with the younger generation.

I am grateful for comments, conversations, correspondence, and/or moral support from several colleagues and friends: Patrick Grim, Ágnes Hárs, Bryan D. Jones, József Lázár, Scott Page, András Schubert, András Simonovits, Zsuzsa Szvetelszky, Ferenc Tátrai, and Jan Tobochnik. Two of my excellent former students at Kalamazoo College kindly helped me. Caroline Skalla prepared high-quality and legally transparent figures, and Raoul Wadhwa copy-edited the original Hunglish text. Big thanks, Caroline and Raoul!

I thank the Kalamazoo College community, specifically my close colleagues, who provided a friendly, intellectual atmosphere. I am also indebted to my colleagues at the Department of Computational Sciences at Wigner Research Centre for Physics in Budapest. I also thank the Henry R. Luce Foundation for letting me serve as a Henry R. Luce Professor.

I almost finished the book when I conversed with Dennis Bernstein in Ann Arbor. He tried to explain his perspective, and I hope I understood it. A single sentence can't grasp his whole approach. Still, it is illuminating: "The goal of feedback control is to achieve the best possible performance through closed-loop uncertainty mitigation, recognizing that performance is the objective of *control*, but uncertainty is the *raison d'être* for feedback *control* [9].

I am thankful to Springer Editors Thomas Ditzinger, who worked with me initially, and Leontina di Cecco, for her professional help in completing this book.

My wife, Csuti's support, love, and wisdom greatly benefited me. It isn't easy to express my gratitude.

Kalamazoo, USA   Péter Érdi

References

1. Èrdi P, Szvetelszky Zs (2022) Repair: When and how to improve broken objects ourselves, and our society. Springer
2. Meadows DH, Meadows DL, Randers J, Behrens WW (1972) The limits to growth: A report for the Club of Rome's project on the predicament of mankind. Earth Island, London
3. Wiener N (1950) The human use of human beings. Houghton Mifflin
4. Arthur WB (1994) Increasing returns and path dependence in the economy. Ann Arbor, MI: University of Michigan Press
5. Diamond J (2004/2011) Collapse: How societies choose to fail or succeed. Viking Press
6. Tetlock PE, Gardner D (2016) Superforecasting: The art and science of prediction. Crown
7. Cavana RY et al. (2021) Feedback economics: Economic modeling with system dynamics. Springer
8. Golledge N (2023) Feedback: Uncovering the hidden connections between life and the Universe. Prometheus

9. Bernstein DS (2022) Facing future challenges in feedback control of aerospace systems through scientific experimentation. J Guidan Control Dynam 45(2202–2010)

Contents

Chapter 1
Norbert Wiener's Dream: Technology, Life, and Society

Abstract This chapter overviews the birth of Cybernetics as a general theory of goal-seeking systems. It introduces the notion of positive and negative feedback and gives credit to the General Systems Theory, which gave a framework for integrating the natural sciences with the social sciences. Modern computers made possible the simulation of social systems based on causal relationships among state variables, leading to various predicted scenarios about the world's future, some pessimistic. One clear insight is that uncontrolled technological progress may negatively affect the environment, and we need to maintain a balance between economic growth and sustainability.

1.1 From Military Technology to Predicting Human Behavior

1.1.1 From Pure Math via Military Applications to a General Theory and World View

1.1.1.1 From Pure Math to Antiaircraft Prediction

Norbert Wiener was a prodigy mathematician. He was eighteen when he received his Ph.D. in the Philosophy of Mathematics at Harvard University. While the basis of his reputation came due to his results in "pure" mathematics (from sub-fields called harmonic analysis and the statistical theory of time series), at MIT he cooperated with electrical engineers on a practical problem. In early wartime, Wiener's interest turned to the problem of destroying enemy airplanes. His goal was to design both, an algorithm and the physical implementation of this algorithm using electric circuits. To help the transition from algorithms to physical implementation, he hired Julien Bigelow (1913–2003), an MIT-trained electrical engineer.

Wiener called the construction the "anticraft (AA) predictor". By applying his deep knowledge of the theory of random time series, his algorithm analyzed the zigzagging motion of enemy planes and made predictions for their future positions. Then, an anti-aircraft shell was launched at the aircraft. Wiener and his colleagues

© The Author(s), under exclusive license to Springer Nature Switzerland AG 2024
P. Érdi, *Feedback*,
https://doi.org/10.1007/978-3-031-62439-1_1

designed a *goal seeking* system. Either the goal was achieved (the plane went down) or not. Interestingly, the classical deterministic algorithm, which did not consider the bomber's irregular motion, was not inferior to Wiener's method.

1.1.2 Towards a General Theory of Goal-Seeking Systems

While Wiener was disappointed with the numerical results, his ambition increased as he saw in the AA predictor a general framework for a new understanding of the *human-machine relationship* [1]. His mind connected the problem of hitting fast, maneuverable bombers with ground-based artillery to the theory of goal-seeking systems. The point is that soldiers, calculators, and firepower were considered parts of a single integrated system, and feedback mechanisms were the key implementation tools.

Wiener soon took more steps as he became interested in the predictor's physiological, psychological, and philosophical consequences. The new approach and its implications for science and technology, in general, was published in 1943 by Arturo Rosenblueth (1900–1970), a Mexican physiologist, Wiener, and Bigelow, and was titled 'Behavior, Purpose, and Teleology'. A hindsight view reveals that the paper is strange in several aspects. It was published in a journal called *Philosophy of Science* [2] and was not a technical paper. Not only did it not contain a single formula or figure, but not even a single reference. In any case, the paper emphasized that **purposeful behavior** can exist both in engineered and biological systems **without** assuming the Aristotelian "final cause". (Aristotle famously defined four causes to answer the question *Why?*: *material*, *formal*, *efficient* and *final* ones [3]) Present causes can explain purposeful behavior, but the causation acts circularly.

Cause and effect

Common sense thinking and problem-solving often adopt the concept of *a single cause and a single effect*. The proverb **you reap what you sow** reflects that previous actions generate future consequences. A cause tells us *why* something happens. An effect tells us *what* happened. We have all heard Newton's story and know that the apple fell because of gravitation.

Circular causality, in essence, is a sequence of causes and effects whereby the explanation of a pattern leads back to the first cause and either confirms or changes that first cause. For example, A causes B, and B causes C, which modifies A. The concept itself had a bad reputation in scientific circles. Arguments and processes that did not have a well-defined end, in other words, if they 'went nowhere,' were described as *vicious circles*. People who could not progress were characterized as "going in circles". In conventional logic, *circular reasoning* was considered pathological and to be avoided. Circular causality was reintroduced

> to science by a then-new field **cybernetics**, which emphasized feedback and self-regulation. In a feedback system, there is no clear discrimination between causes and effects since the output influences the input.

Wiener and his colleagues analyzed the concept of purpose and realized that not *all* machines are purposeful. As a mechanical device, a roulette wheel was designed to be purposeless. A clock has a purpose but not a *final state* toward which the clock's movement tends. What about a gun? It may be used for a purpose, but the goal is not a built-in feature of the gun. In contrast, a torpedo with a target-seeking mechanism is intrinsically purposeful. The key element of this mechanism is **feedback control**.

Engineers understood the general principles of feedback control, and autonomous control systems were used to replace human operators to eliminate errors due to human failure. It turned out that living systems also use feedback control. Rosenblueth worked with the Harvard physiologist Walter Cannon (1871–1945), who elaborated the theory of "homeostasis", and popularized it in the book *The Wisdom of the Body* [4]. Modern physiology considers living processes as *self-regulated*. As we already know, Wiener and Bigelow were involved in developing anti-aircraft guns during the Second World War by using negative feedback control. So, the cooperation of three authors, a physiologist, a mathematician, and an engineer, was very reasonable and proved fruitful.

Feedback is defined as a process whereby some proportion of a system's output signal affects (i.e., feeds back) the input. Therefore, the system itself contains a loop. Feedback mechanisms (or lack thereof!) fundamentally influence the dynamic behavior of a system. This behavior might lead to a resting state and periodic or irregular volatile motion. Explosion and extinction are phenomena occurring in the lack of stabilizing feedback mechanisms. Roughly speaking, negative feedback reduces the error or deviation from a goal state and has stabilizing effects. Positive feedback, which increases the deviation from an initial state, has destabilizing effects. As we know, natural, technological, and social systems are full of feedback mechanisms (Fig. 1.1).

On the topic of social scientists, the then-married and later very influential anthropologist couple, Margaret Mead (1901–1978) and Gregory Bateson (1904–1980), were both enthusiastic about seeing the omnipresence of feedback mechanisms. In a fascinating interview [8], they recollect their memories about the birth of cybernetics.

Mead: "Now, there were some other things like this that were being talked about, and one was what was called a vicious circle. Milton Erickson had written a paper on a girl who quarreled and had headaches and got alienated from people, which led to further quarrels, and so on."

Bateson: "Yes, all the positive feedback stuff was ready. And that presented the problem: why don't these systems blow their tops? And the moment they came out with negative feedback, then one was able to say why they don't blow their tops. …"

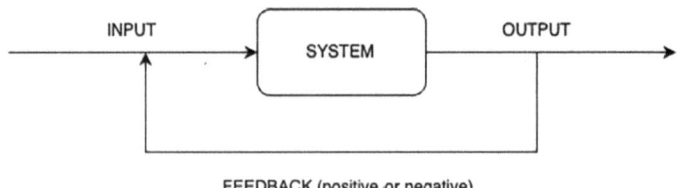

Fig. 1.1 Systems with feedback

But nobody put the stuff together to say these are the formal relations that go for natural selection, which go for internal physiology, which goes for purpose, which goes for a cat trying to catch a mouse, which go for me picking up the salt cellar" [8].

When flushed, the toilet uses *negative* feedback to fill itself with water. In this case, an impairment of the control system may show positive feedback, which implies a non-required overfill. Another simple example of negative feedback is a connected thermostat-heater system. A sensor detects the temperature, and when the temperature reaches a predetermined value, the thermostat signals the furnace to switch off. When the temperature drops below another predetermined value, the furnace is turned back on. This feedback loop has been further enhanced in modern thermostats, which incorporate time delays and predictions about the temperature to perform more smoothly and avoid too many swings and cycles [5].

If a system responds to perturbation in the same direction as the perturbation and amplifies the initial minor deviations, it exhibits *positive* feedback. The uncompensated, boundless positive feedback frequently leads to extreme events, extinctions, explosions, and other disasters, as we will see in later chapters of this book.

Please note that the adjectives "negative" and "positive" refer not to desirability but to the direction of feedback relative to the input. This is distinct from their meaning in educational psychology, in which positive reinforcement (or feedback) is adopted to incentivize subjects to continue a behavior.

1.2 Cybernetics: Heritage and Revival

Cybernetics, as a scientific discipline, has been named by Wiener in his book "Cybernetics", with the subtitle "Control and Communication in the Animal and the Machine" [6]. While physiologists already knew that the involuntary (autonomic) nervous systems control Bernard's "internal milieu" (to be discussed later in 3.2.2), Wiener extended the concept by suggesting that the voluntary (somatic) nervous system may also control the environment via feedback mechanisms. The theory of goal-oriented behavior based on self-regulating mechanisms (also called servomechanisms) promised a new framework to understand the behavior of animals, humans, and computers just under design and construction at that time.

1.2.1 The Cybernetic Movement

There was a series of conferences (1946–1953) (sponsored by and named after the Macy Foundation–often erroneously believed to be related to the department store) in which Norbert Wiener also played an important role. Cybernetics was American and soon became also British. It was labeled (together with genetics) as bourgeois pseudoscience in Stalin's Soviet Union. (I find the coincidence that there was only several days' difference between Churchill's Iron Curtain speech in Fulton and the first Macy conference on cybernetics (March 5th, March 8–9th, 1946) remarkable.) The last conference was held several weeks after Stalin's death. Interestingly, but not very surprisingly, after the decline of cybernetics in the U.S., it became popular in the Soviet scientific community. Maybe it is not literally true that cybernetics became a suspicious word in the US, but some people say, "Well, it is nothing else but computer science"; others somehow identify it with robotics).

The Macy conference series was organized to understand "Feedback Mechanisms and Circular Causal Systems in Biological and Social Systems," as the title of the first conference suggested. (Wiener's term **cybernetics** appeared only later.) The conferences had an interdisciplinary character.

While "cyberneticians" partly spoke on behalf of physics (well, a strange physics, not a physics of matter and energy, but a physics of information, program, code, communication, and control), there was no professional physicist among them. Max Delbrück (1906–1981), (who was trained as a physicist but had already transitioned to applying physics principles to biology) was invited since the very influential John von Neumann (1903–1957) felt that molecular genetics would be interesting from a mathematical perspective. Delbrück did not like the conference and never returned. As the French philosopher Jean-Pierre Dupuy [7] analyzes, it is one of the most striking ironies in the history of science that the significant attempt of molecular biology to reduce biology to physics happened by using the vocabulary of the cyberneticians. "Cybernetics, it seems, has been condemned to enjoy only posthumous revenge" [7], pp. 78.

Wiener and von Neumann, in particular, claimed that their theories and models would be helpful in economics and political science.

The main topics of the conferences were [7]:

- Applicability of a Logic Machine Model to both Brain and Computer
- Analogies between Organisms and Machines
- Information Theory
- Neuroses and Pathology of Mental Life
- Human and Social Communication

There is no doubt that cybernetics was an intellectually appealing and ambitious discipline and was partially a victim of its own ambition. However, many of its tenets survived under the names of other disciplines, and I think cybernetics now strikes back. But first, we should remember the chairman of the Macy conferences, Warren McCulloch.

1.2.1.1 McCulloch: A Pioneer of Interdisciplinarity

In addition to Wiener, McCulloch was the other Founding Father of the movement and scientific discipline of cybernetics. He had a particular personality, was very creative, and was a polymath. He learned philosophy, became a physician, and trained in mathematical physics and physiological psychology. McCulloch was an experimentalist, a theoretician, a premodern scientist, a philosopher, and possibly a magician. (For a very authentic review of McCulloch's personality, life, and works, see the short biography of Michael Arbib [10], one of McCulloch's last students).

Between 1941 and 1952 (i.e., in the initial period and during the golden age of cybernetics), he was in Chicago at the Neuropsychiatric Institute of the University of Illinois in Chicago. Then, he moved to the Department of Electrical Engineering at MIT to work on brain circuits. However, the department's abbreviation EE had a different meaning to him. McCulloch founded a new field of study based on this physical and philosophical intersection. He called this field of study "Experimental Epistemology," the study of knowledge through neurophysiology. In the same year–1943–when the "Behavior, Purpose and Teleology" was published, McCulloch and the prodigy Walter Pitts (1926–1969) published a paper with the title "A Logical Calculus of the Ideas Immanent in Nervous System," which was probably the first experiment to describe the operation of the brain in terms of interacting neurons [9].

The MCP model was established to capture the logical structure of the nervous system. Therefore, cellular physiological facts known even at that time were intentionally neglected. A massive industry called Artificial Neural Networks grew from this model. After many ups and downs, "deep learning in neural networks" [11, 12] has become a celebrated sub-field of today's machine learning / artificial intelligence discipline. (We will discuss the potential risk of modern AI at 6.1.1).

Cybernetics in the Soviet Union

Historically, it is remarkable to see that before 1955, cybernetics was anathema and, at most, a "bourgeois pseudoscience" and soon became a slogan in the development of Soviet science and technology. In 1956, at the Twentieth Congress of the Communist Party of the Soviet Union, Khrushchev denounced Stalin. Soon, industrial-process automation became a goal, and the Scientific Council of Cybernetics was established. Andrey Kolmogorov (1903–1987), the celebrated (informal) Moscow mathematics school leader, wrote a laudatory article about cybernetics. Michael Arbib, then a young cybernetician, visited the Soviet Union in 1964 and wrote: "Cybernetics was subsequently so popularized on press, radios, and television, that when I visited the Soviet Union, everybody I met (including a pianist, a customs official, Intourist guides, and hotel staff) knew of cybernetics, and commented on my luck in working in such a new and exciting field–a change from the blank looks the word "cybernetics" calls forth in the West" [13].

1.2.2 General Systems Theory: Bridge Between Cybernetics and Social Sciences

It is easy to conceive that the movement of cybernetics was driven, at least implicitly, by the grand utopia that Metaphysics, Logic, Psychology, and Technology can be synthesized into a unified framework. In a second book [14] entitled *The Human Use of Human Beings* [14], Wiener was quite concerned about the socio-economic aspects of industry automation. He predicted the beneficial elements of automation on society and argued about the need for long-term planning for the next Industrial Revolution. While the very influential Margaret Mead asked Wiener to extend cybernetics to social systems, Wiener felt it would be difficult to make progress due to the lack of reliable data.

Somewhat parallel to cybernetics, another movement, general systems theory, emerged: the Society for General Systems Research, founded in 1954 by the biologist Ludwig von Bertalanffy, who came from Vienna to the USA. Von Bertalanffy considered the closed-loop self-regulatory system perspective of cybernetics as a particular case of general systems, including self-organization in open systems.

Self-organization is not a well-defined concept. Very loosely speaking, *organization* is a process when elements, say people, behave by following external instructions. If a gate change is announced at an airport, a subset of people starts to walk or run, say from gate G2 to H41. Self-organization happens when the elements interact with their local neighbors only and are not directed by any "boss". Spontaneous formation of spatial, temporal, and spatiotemporal structures happens in systems composed of few or many components. Many examples occur in physical, chemical, and biological systems, from forming hexagonal patterns in fluids via chemical oscillations and waves to biological clocks, spatial patterns of animal coats, and collective behavior of bird flocks, insects, and fish colonies. The process of self-organization can be found in many fields, such as economics (growth, competition, extinction of companies) and social psychology (formation of public opinion), among others.

Examples for self-organization: [15]

In *chemical kinetics*, self-organized patterns occur due to the interaction of autocatalytic (i.e., positive feedback) compensated by the reaction step blocking unbounded growth. Oscillatory patterns and ordered spatial structures (blobs, stripes) are formed.

Population dynamics, ecological systems: connectivity, stability, diversity, resilience, resistance. The fundamental question is how the stability of an ecological system changes if there is a change in the connectivity of the network of interacting populations and/or in the strength of the interactions. Model studies show that weak connections enhance stability, and they may be the glue that binds natural communities together.

Epidemics are characterized by a rapid increase in the size of the infected population due to the interaction between infected and susceptible individuals. The infected sub-population can spontaneously be converted to a removed pool or by external control.

Segregation dynamics. Local rules (micro-motives) imply globally ordered social structures (macro-behavior). Simulation results suggested that a slight preference to live independently implies global segregation and the formation of ghettos.

Opinion dynamics. Interaction of people in a group implies changes in their opinions about different issues may lead to consensus, fragmentation, and polarization.

Business cycles. In a very influential model, Nicolas Kaldor (1908–1986) provided a mechanism for generating temporal oscillatory dynamics in income and capital by assuming a nonlinear dependence of investment and saving on income.

Self-organization in the nervous system. According to embryological, anatomical, and physiological studies, the wiring of neural networks results from the interplay of purely deterministic (genetically regulated) and random (or highly complex) mechanisms. Fluctuations may operate as an "organizing force." Self-organizing developmental mechanisms (considered pattern formation by learning) are responsible for the formation and plasticity of ordered neural structures. Evolvability, the basis of self-organization, poses constraints on brain dynamics. A stable internal representation of the external world indicates the presence of attractors. Here, an attractor refers to a state in a system toward which the system naturally settles after starting from a given initial condition. Self-organization needs these attractors to have sufficient instability to be able to alter to adapt to the environment.

Systems theory is interested in **similarities** and **isomorphism**, not in the differences of various systems. The basic assumption is that the same fundamental principles govern physical, chemical, biological, and psychological systems.

Bertalanffy's General System Theory:

(1) There is a general tendency towards integration in the various sciences, both natural and social.

(2) Such integration seems centered on a general systems theory.

---

(3) Such theory may be an important means of aiming at exact theory in the nonphysical fields of science.

(4) By developing unifying principles that run 'vertically' through the universe of individual sciences, this theory brings us closer to the goal of the unity of science.

(5) This can lead to a much-needed integration into scientific education.

Of course, we know the collective wisdom: What's the difference between generalists and specialists? A generalist knows less and less about more and more until, eventually, he or she knows nothing about everything. A specialist knows more and more about less and less until, eventually, he or she knows everything about nothing. Being either a generalist or a specialist is useless, and anyone trying to be both at the same time inevitably self-destructs [16].

1.2.3 From Cybernetics via System Dynamics to Limits to Growth

Causal influences among variables determine the dynamics of any system [18]. Figure 1.2 gives some self-explanatory examples of elementary causal interactions among two variables:

How do we interpret these diagrams? We may assume that *population* denotes the number of rabbits per square kilometer of a forest, while *birth* is a process reflecting the birth rate of rabbits. By using the terminology of *system dynamics*, they are **stock** and **flow** variables. The two positive signs express the quality of the interaction: a more significant birth rate implies a larger population, and a larger population induces a more significant birth rate. Please note that you see a simple causal loop between two variables. Everything else, such as the death rate, immigration, and emigration, is neglected. But you can build a large model by using many simple causal loops.

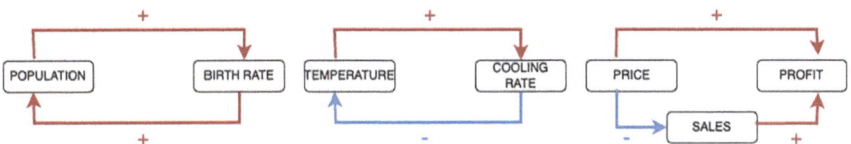

Fig. 1.2 Simple causal loops: mutual activation; activation and inhibition; inspired from [18]

1.2.3.1   The Route Towards the Computer Simulation of Social Systems

The man who understood the role of causal loops in the operation of complex systems and elaborated the theory and software (with coworkers, of course) to analyze their possible behaviors was Jay Wright Forrester (1918–2016). He was involved in implementing servomechanisms of an alert and defense system to detect and prevent a potential attack on American territory by Soviet rockets. Forrester adopted Wiener's concept of applying the control of complex organizations involving men and machines in *real time* [19]. It is somewhat unbelievable that Forrester also patented the magnetic-core random-access memory to make real-time computation possible. As he moved from developing hardware to building a School of Management at MIT, he realized that unexpected dynamic phenomena, such as undesired oscillatory effects in business life, resulted from the interaction of feedback loops. In his book Industrial Dynamics [20], he combined the feedback concept with computer simulation, as the new computers at MIT made it possible to study the dynamic behavior of larger-scale systems. The field of System Dynamics was born. The next significant step was a transition from modeling industrial and management problems to *public policy* when Forrester cooperated with the former mayor of Boston, John Collins (1919- 1995), to investigate the causal mechanisms behind the different stages of urban dynamics, such as development, stagnation, decline, and recovery [21]. Forrester understood that computational models should be superior to simple debates about potential social policies. Computer simulations (i) generate insight into the causes of problems and (ii) help understand the likely effects of proposed solutions. He argued that due to our cognitive limits, our qualitative insight is often misleading, and computer simulations may result in counter-intuitive surprising behavior of social systems. In 1970, Forrester was invited by the Club of Rome, an organization founded in 1968, to study the problems that plagued our species; this was likely in response to major events such as the assassinations of Martin Luther King (1929–1968) and Robert Kennedy (1925–1968), the protests and rebellions in Paris and Chicago, and the presence of Soviet tanks in Prague. The question was what system dynamics could teach us about the dynamics of global development. Forrester's first decision was that the state of the global world could be characterized by a few variables, such as population, natural resources, pollution, agricultural and industrial production, capital investment, and quality of life. The variables influence each other, reflected by causal loops. Our mind can not draw the consequences of the chain or network of simple causal loops, say, an increase in industrial production implies an increase in population and pollution. An increase in pollution has a negative effect both on food production and directly on the population. Complex world problems should thus be addressed with the aid of computational modeling.

If you make simulations, you could and should give different numerical values of the parameters reflecting the strength of interactions. Forrester made several simulations and published the book World Dynamics [22]. For a not-very-painful technical explanation for the basic versions of the World Model, see [23]. This book was the basis of the work of Forrester's student Dennis H. Meadows and his team that led to the publication of the mega-bestseller The Limits to Growth [24]. It presented 12 scenarios for the works from 1970 to the year 2100. Six were pessimistic (too many people, too few resources, too little food, or too much pollution). Six scenarios did not show overshot or collapse but what we now call sustainability. The book's main conclusion is that 1970 economic growth, resource use, and pollution rates continued unchanged, and the world would face economic and environmental collapse in the mid-twenty-first century. But the book also called for action. Changing the growth rates and other parameters to gear the world toward stability and sustainability was possible. The book was a warning signal that the future is in our hands and the starting point of extensive discussion about the interaction of economic growth and ecological effects.

1.2.3.2 Loud Feedback on Silent Spring

Literally, *Limits to Growth* was not the starting point. Famously, Rachel Carson published Silent Spring [25] in 1962. She documented that using the pesticide DDT had caused damage to agricultural animals, bees, birds, domestic pets, wildlife, and even humans [26]. The book was a passionate warning about the future of the Earth and all life on this planet. It was also a loud call for action as it criticized uncontrolled technological progress, eventually inspiring the modern environmental movement. Let us take one step back to demonstrate the power of unintended consequences and the necessity of balanced thinking. Paul Hermann Müller (1899–1965) was a Swiss chemist who discovered in 1939 that DDT could kill insects and proved effective against the Colorado potato beetle. Further studies demonstrated its incredible effectiveness against various pests, including fleas, louse, mosquitoes, and sandflies, which spread malaria, typhus, yellow fever, the plague, and various tropical diseases, respectively. Müller was motivated by a significant food shortage in Switzerland, which underscored the need for a better way to control insect infestation of crops. DDT played a positive role in the Second World War by clearing South Pacific islands of malaria-causing insects for U.S. troops. Another motivation that Müller had was the lethal typhus epidemic in Russia. It is difficult to deny that DDT saved millions of lives and was helpful against malaria. However, it is also difficult to deny the unintended long-term consequences. Müller cannot be blamed for not seeing the long-term implications of his discovery.

The use of pesticides after the finish of the war increased in agriculture. Carson documented that DDT entered the food chain, accumulated in the fatty tissues of animals and humans, and caused cancer and genetic damage. The book describes an imaginary American city: "The roadsides, once so attractive, were now lined with brown and withered vegetation as though swept by fire. These, too, were silent, deserted by all living things. Even the streams were now lifeless. Anglers no longer visited them, for all the fish had died. ...No witchcraft, no enemy action had silenced the rebirth of new life in this stricken world. The people had done it themselves."

Mainstream economists viciously attacked The Limits to Growth, and later, U.S. President Ronald Reagan famously declared that "There are no great limits to growth when men and women are free to follow their dreams." Parallel to the Limits to

Growth, another bestseller *Only one Earth: the care and maintenance of a small plane* was published as an unofficial report commissioned by the Secretary-General of the United Nations. Conference on the Human Environment [27] also contributed to the dramatic increase in ecological awareness.

As a compromise between economic growth and concerns about resources and pollution, the concept of *sustainable development* was suggested by the Brundtland Report in 1987 [28]. The Rio Process was initiated at the 1992 Earth Summit in 2015, and the United Nations General Assembly adopted the Sustainable Development Goals (2015 to 2030). The concept of sustainability is normative; it suggests how a system *should* work. However, it should be supplemented with a mechanism for implementation. Controlling sustainability needs a systematic, interdisciplinary study to find the balance between non-zero economic growth and resource management.

"Climate action is just one form of sustainability — a concept that asks us to see the big picture of what we are taking from the Earth and how our actions affect future generations" [29]. Mechanisms of global warming generally contain a positive feedback loop when warming increases the amount of water vapor in the atmosphere, leading to further warming. It also includes a negative feedback loop, as more energy will radiate out. (Readers with technical backgrounds know that the total energy radiated per unit surface area of a black body is proportional to the fourth power of the black body's temperature–known as the Stefan-Boltzmann law). Small changes in the parameters of these feedback loops may have dramatic effects. Understanding and quantifying the impact of such feedback mechanisms are among the top priorities of current climate-change research.

1.3   Lessons Learned and the Book in Prospect

A new scientific field around the concept of feedback emerged from the military problem of designing specific antiaircraft systems. Norbert Wiener and Warren McCulloch were the founding fathers of Cybernetics. As Wiener's book subtitle suggested, a unified theory of animals and machines might exist. While such nobilities of social sciences as Margaret Mead and Gregory Bateson were enthusiastic participants of the early cybernetic conferences, there were neither sufficient data nor sophisticated methods for analyzing social systems. In the nineteen-sixties, the analysis of large-scale systems based on the concept of causal loops and using MIT's new computational tools became possible.

The 1950s in the US seemed happy after the decades of the Great Depression and World War II, as homes became affordable and the suburban population exploded. They were somewhat happy days, but not for all: the US was still very segregated. Consumerism increased and implied the rise of packaging and food waste; industrial growth led to a dramatic increase in pollution. Warning signals induced environmentalism, and humankind's contradiction between growth and sustainability became a central theme.

This book is about the possible beneficial role of feedback in avoiding catastrophic scenarios and keeping the world on sustainable trajectories. Chapters 2 to 5 illustrate how the combination of positive and negative feedback loops ensures stable, "normal" operation and how the impairment of the control systems leads to pathological and malignant performance. Chapters 6 and 7 consider the stabilizing and destabilizing mechanisms. The key is that the combination of building early warning systems with decision-making and actors is our best hope to avoid catastrophic scenarios.

References

1. Galison P (1994) The ontology of the enemy: Norbert Wiener and the cybernetic vision. Crit. Inquiry 21(1):228–266
2. Rosenblueth A, Wiener N, Bigelow J (1943) Behavior, purpose and teleology. Phil Sci 10(1)
3. Falcon A (2022) Aristotle on causality. In: Edward NZ (ed) The stanford encyclopedia of philosophy, Springer https://plato.stanford.edu/archives/spr2022/entries/aristotle-causality
4. Cannon WB (1963) The wisdom of the body, 2nd edn. W. W, Norton and Company, New York
5. Aström KJ, Murray RM (2012) Feedback systems. Princeton University Press
6. Wiener N (1948) Cybernetics: Or control and communication in the animal and the machine. MIT Press, Paris, Hermann & Cie & Camb. Mass
7. Dupuy JP (2001) The mechanization of the mind: On the origins of cognitive science. Princetion University Press
8. Brand S (1976) For god's sake, Margaret! Conversation with Gregory Bateson and Margaret Mead coevolutionary quarterly
9. McCulloch WS, Pitts W (1943) A logical calculus of the ideas immanent in nervous system. Bull Math Biophys 5:115–133
10. Arbib MA (2000) Warren McCulloch's search for the logic of the nervous system. Perspect Biol Med 43(193–216)
11. LeCun Y, Bengio Y, Hinton G (2015) Deep learning. Nature 521:436–444. https://doi.org/10.1038/nature14539
12. Schmidhuber J (2015) Deep learning in neural networks: An overview. Neural Netw 61:85–117. arXiv:1404.7828, https://doi.org/10.1016/j.neunet.2014.09.003
13. Arbib M (1966) A partial survey of cybernetics in Eastern Europe and the Soviet Union. Behav Sci 1:193–216
14. Wiener N (1950) The human use of human beings. Houghton Mifflin (US) Eyre & Spottiswoode (UK)
15. Érdi P (2007) Complexity explained. Springer
16. https://www.primegenesis.com/our-blog/2010/09/the-difference-between-a-generalist-and-a-specialist/
17. Conway F, Siegelman J (2005) Dark hero of the information age: In search of Norbert Wiener, the father of cybernetics. Basic books
18. http://systems-sciences.uni-graz.at/etextbook/sw1/causal_loops.html
19. Lane DC (2007) The power of the bond between cause and effect: Jay Wright Forrester and the field of system dynamics. Syst Dyn Rev 23(2–3):95–118
20. Forrester JW (1961) Industrial dynamics. Pegasus Communications, Waltham, MA
21. JW Forrester (1969) Urban dynamics. Pegasus Communications
22. Forrester JW (1971) World dynamics. Wright-Allen Press, Cambridge, MA
23. Castro R (2016) Overview of the Modelica-based World2 and World3 models. MOSES. https://openmodelica.org/images/M_images/Moses206/moses2016-25-RodrigoCastro-Overview-of-the-Modelica-based-World2-and-Wold3-models.pdf
24. Meadows DH, Randers J, Meadows DL (1972) The limits to growth. Universe Publ

25. Carson R (1962) Silent spring. Houghton Mifflin Company
26. https://www.nrdc.org/stories/story-silent-spring
27. Ward B, Dubois R (1972) Only one earth. The care and maintenance of a small planet, W. W, Norton, New York
28. United Nations General Assembly (1987) Report of the world commission on environment and development: Our common future
29. What is "sustainability"? Is it the same thing as taking action on climate change? MIT Climate Portal, March 30, 2021

Chapter 2
Feedback Control in the History of Technology

Abstract This chapter gives an overview of the application of feedback control throughout technological development. In technological systems, the goal is to ensure that some physical quantities, such as temperature, pressure, velocity, or altitude (and many others), show some desirable, prescribed behavior over time. Negative feedback stabilizes, while positive feedback amplifies even the initially minor differences. Outriggers, mechanical clocks, steam engines, aviation technologies, and electronics are the main stages of progress toward the Space Travel Age. Feedback control systems played a crucial role in space exploration.

2.1 Feedback Control: A Modern Concept is a (Not So) Invisible Thread in the History of Technology

2.1.1 Control is Everywhere

A control system is a device or process that modifies the behavior of another device or system to achieve a specific goal. Control systems are split into two major classes: (i) natural (primarily biological) and (ii) man-made control systems. We use the latter in our daily life. Humans adopt a two-stage strategy to construct them. A device to achieve a specific goal is first, *designed*, and second, *manufactured*. It is possible to control processes without evaluating their output behavior. This is called **open loop** control. It does not measure the condition of its output signal, as there is no feedback. The textbook example is a *central heating boiler* controlled only by a *timer*. The process variable is the building temperature. This controller operates the heating system for a constant time *regardless* of the temperature of the building. The control action is the switching on or switching off of the boiler.

While feedback control, which naturally implements **closed loop** control, is an engineering discipline, it is still much older than humanity. In biology, feedback control mechanisms were identified at different levels of hierarchical organization (molecule, cell, organism, or population) that influence the system's continued activity or productivity. Chapter 3 discusses the role of feedback control in providing normal physiological performance. Furthermore, we will study how the impairment of

© The Author(s), under exclusive license to Springer Nature Switzerland AG 2024

P. Érdi, *Feedback*,

https://doi.org/10.1007/978-3-031-62439-1_2

15

the control system leads to what is called *dynamical diseases*. A better understanding of pathological mechanisms may give the hope to offer new therapeutic strategies.

Feedback control is a modern concept. As Dennis Bernstein, a professor of Aerospace Engineering at the University of Michigan, taught us in an excellent paper, it is an invisible thread in the history of technology, as studies on mechanical clocks, steam engines, aerodynamics, and electronic devices show [1]. As such, its progress is closely tied to the practical problems that must be solved during any period of human history.

In technological systems, the goal is to ensure that selected physical quantities, such as temperature, pressure, velocity, or altitude (and many others), show some desirable, prescribed behavior over time. For example, cruise control in a car maintains constant speed independent of road inclines; similarly, the autopilot maintains speed, altitude, and heading. This is an example of a more sophisticated automatic control system that works without requiring constant manual control by a human operator. An autopilot system is sometimes colloquially referred to as *George*. (e.g., "we'll let George fly for a while"). Some sources suggest that it relates to inventor George De Beeson (1897–1965), who patented an autopilot in the 1930s. According to another story, the Royal Air Force pilots coined the term during World War II to symbolize that their aircraft technically belonged to King George V. While we do not necessarily need to know the answer (see [2]), it is interesting to read an excerpt from George De Beeson's patent application:

> *This invention relates to airplanes and, more especially, to automatically fly the machine substantially without manually operating several control surfaces. The invention aims to provide a simple, reliable, and effective apparatus for automatically restoring the airplane to a given and normal longitudinal and traverse balance if it tips or pitches to an undesired angle from the normal.*

We apply control in our everyday biological and social lives. For instance, when we walk, grasp an object, suppress our reflex to angrily respond to our boss (it happened to me today, February 2nd, 2023), or drive a car. How do we maintain the speed of a car on the highway? We get some input data by observing the speedometer and then comparing the data with a number prescribed by the law. In the next step, we take action and increase or decrease the pressure on the gas pedal. The process of controlling the speed can be transferred from human to machine, so we adopt an automatic speed control system. This system contains an algorithm based on a mathematical model to make decisions and take action. Information about the actual speed is fed back to the controller by sensors, and the control decisions are implemented via a device called the *actuator*, which increases or decreases the fuel flow to the engine.

Otto Mayr, a legendary historian of technology and a former director of the National Museum of History and Technology in Washington DC and the Deutsches

Museum in Munich, defined three criteria for determining whether a device is a feedback mechanism [3, 4].

1. The purpose of a feedback control system is to carry out commands; the system maintains the controlled variable equal to the command signal despite external disturbances.
2. The system operates as a closed loop with negative feedback.
3. The system includes a sensing element and a comparator, at least one of which can be distinguished as a physically separate element.

While Mayr believed an ancient Greek water clock was the first device to fulfill these criteria, a feedback device was adopted much earlier in Polynesia.

2.1.2 A Prehistoric Feedback Mechanism: The Polynesian Outrigger Canoe

The Disney film Moana got two Oscar nominations in 2017. The title character and her companion, the demigod Maui, sail a Polynesian *double-hulled outrigger* canoe. Throughout the film, Moana learns to navigate the ocean without a compass or map. Moana and Maui have an incredible knowledge of the stars, the horizon, and the look and feel of ocean swell (Moana's song *How Far I'll Go*, see [6]). We know that Polynesian sailors could set a safe course toward their destination. The adopted technological solution for increasing the roll stability of the boats by adopting feedback is remarkable. It is quite possibly the first feedback mechanism created by humanity, predating the float valve by at least a millennium, as an exciting paper demonstrated [5].

2.1.2.1 The Outrigger is a Stabilizing Device Implemented by a Feedback Mechanism

Any watercraft with poor roll stability will likely capsize. The so-called outrigger boats have *lateral* support floats known as outriggers, fastened to one or both sides of the main hull. The role of an outrigger is two-fold. First, the outrigger adds buoyancy to the vessel since the outriggers are made of materials that float irrespective of their shapes. Second, and more importantly, adding a float at the end of a boom dramatically increases the **roll stability** of small canoes.

Both single-hulled and double-hull boats have been constructed. In the latter, an outrigger generates stability due to the distance between its hulls rather than the shape of each hull. As such, the hulls of outriggers are typically longer, narrower, and more hydrodynamically efficient than those of single-hull vessels.

When the canoe rolls to raise the float from the water, the weight of the float at the end of a moment arm provides a moment that rotates the float back to the surface. When the rotation of the canoe acts to push the float into the water, the float's buoyancy restores the float to the water's surface. More technically, the outrigger *senses* and resists angular disturbances, making it a prototype of a negative feedback mechanism, and fulfills all the three criteria of Mayr's definition [5].

---

(1) The outrigger **enhances the roll stability** of the canoe. That is, the outrigger helps prevent the canoe from capsizing. In this respect, the regulated variable is the angle between the bottom of the canoe and the water's surface. The outrigger performs this function in the presence of internal disturbances (movement within the boat) or external (waves and wind).

(2) The outrigger provides **negative feedback**. From a nominal position on the water, the buoyancy of the outrigger resists rotations that tend to submerge it during the weight The outrigger resists rotations that tend to raise it out of the water.

(3) The outrigger is a **separate element** from the canoe. The outrigger is the sensor, feedback algorithm, and actuator for detecting and correcting rotations. Its only purpose is to enhance roll stability and disturbance rejection.

The outrigger technology made the Polynesian migration possible, as Austronesian people crossed the Pacific and Indian Oceans. The reconstruction of the story is complicated because of the lack of written language and knowing that metalworking was not used. It was estimated that the outrigger feedback mechanism precedes the water clock by at least 1200 years. In addition, while the first Greek water clock did not have a big effect on society, the Polynesians combined their technology with their remarkable ability to navigate by using stars and navigating birds and connecting surprisingly distant parts of the Polynesian Archipelago.

2.2   From Water Clocks to Mechanical Clocks

2.2.1   The Ancient Greek Water Clock

2.2.1.1   The Technological Invention

Measuring the passage of time was first related to the apparent motion of celestial bodies. A **calendar** system gives names to hierarchical time units. While years, months, and days have a natural (physical, biological) basis, the week cycle seems to have a cultural origin. **Clocks** typically measure time intervals within a day. Sundials were calibrated based on the direction of shadows; gradually melting wax also measured the continuous passing time.

Water clocks (Klepsydra), one of the oldest time-measuring instruments, were used in many cultures. Ordinary water timers were adopted; the passing time was assumed to be proportional to the level dropped. There were some problems with the water clock. The first problem was that a constant water pressure was needed to keep the water flow steady. The second was the necessity of water clocks to match sundials.

As the water level dropped, the water came out more slowly, and the measurement procedure contained a systematic error. Ctesibius (or Ktesibios, 285-222 BCE) worked in his father's barbershop in Alexandria. The constant dripping of water inspired his invention. Ktesibios saw the solution as a more precise measuring device: if the vessel were always full, the water pressure outflow would always be the same.

Ktesibios tried to obtain a uniform inflow of water; the challenge was to keep Klepsydra's reservoir full at all times. First, he added another water tank above the main reservoir. If the reservoir were always full, the water would always come out at the same velocity. He solved this problem. Next, he needed a measuring device. He put another water tank under the constant outflow. In this container, he placed a *float* with a pointer on top and a scale next to it. When the water level rose, the pointer rose at a constant speed. He produced the first mechanical clock and used feedback control. Ktesibios invented one of the first well-documented artificial automatic self-regulatory systems in history by employing a principle to recycle itself automatically. Philon of Byzantium (also known as Philo Mechanicus) used a similar *float regulator* mechanism to keep a constant level of oil in a lamp.

2.2.1.2 Effects on Society

Water clocks were developed, among other purposes, for legal use in Athenian courtrooms to measure the time of the lawyers' and witnesses' speeches; when the water supply ran entirely out of the vessel, the speaker ran out of time.

Klepsydras impacted the ancient world because they increased the importance of the concept of time. These devices provided a better, more efficient way to keep time. Citizens could track what time the markets opened and closed, what time court cases began, during religious sacrifices, for military affairs, and so on. After the original Klepsydra was developed, it impacted many other civilizations. Some adapted it and tried to improve its technology, while others used it daily.

How did the water clock impact later civilizations, including the modern world?

Water clocks significantly impacted the contemporary world by creating a way to keep time. Time is the basis for every aspect of modern culture. Generally, we follow a daily schedule: waking up at a particular time, eating lunch and dinner at approximately the same hour, etc. The impact of the water clock is one of the most significant technological leaps of all time. Without a reliable way to measure time, society would be lost. There would not be a daily schedule or structure. The lives of individuals and institutions are organized around time. We learned at seven years of age that school starts at 8 a.m. Although many years have passed since I attended elementary school, beginning the working day at 8 a.m. is still crucial (I must admit, it has recently become closer to 9 am).

2.2.2 The Mechanical Clock and the Clockwork Universe

In the second half of the 13th century, a new technology appeared in England, France, and Italy: the mechanical tower clock. It showed time independently of the season, the lengths of days, etc. However, mechanical clocks also had disadvantages since they were heavy and weight-driven. Also, they were large, expensive, and initially somewhat inaccurate. The fundamental innovative element in all mechanical clocks is the appearance of a new *regulator*, a complicated mechanism called the **escapement**.

The first escapement was the verge and foliot mechanism. The foliot is a horizontal bar with weights on either end. It sits on a vertical rod called a verge. The verge has pallets to engage and release the main gear. The clocks had two achieve two sub-goals. First, the verge-and-foliot escapement discretized the seemingly continuous time into intervals defined by the duration between impacts. Second, the impacts (ticks and tocks) were counted. By adopting modern terminology, the mechanical clock is a discrete-event system whose dynamics are continuous between impacts and discontinuous at implications.

The Chinese clock is probably not the missing link

Chinese water-clock making was far ahead of Europe's in the 11th century. Joseph Needham (1900–1995), a legendary British biochemist deeply interested in Chinese scientific history, suggested [7, 8] that the escapement was invented in China, already in the eighth century and could be considered as the "missing link" in the technological evolution of clocks between waterclocks and the European mechanical clock. However, the principle behind the European escapement, based on the centrifugal force of an oscillating inert mass, seems unrelated to the Chinese implementation. David Landes (1924–2013), in his excellent book about the revolutionary role of clocks in making the modern world [9], labels the Chinese contribution as a "magnificent dead-end".

The interaction of its components determined the dynamics of the verge-and-foliot. The foliot balance was used for tower clocks. Until the middle of the 17th century, it was also used for weight-driven lantern clocks in people's homes and early and large portable pocket watches exceeding 10 cm in diameter. The next question was how to implement a mechanism where the periodicity comes from an external source and is not determined by the interaction of the components.

The significant development relates to the concept of **isochronism**. The term has a Greek origin (iso + chronos) and means "simultaneously". Clocks are machines for counting oscillator swings. An oscillator, such as a pendulum or balance, has a period independent of oscillator amplitude. A pendulum or balance that takes the same time to complete a swing, no matter how big the swing is, has the property of isochronism. An isochronal oscillator, whether a balance and balance spring, a pendulum, or a quartz crystal in a quartz watch, is essential—without one, there is no timekeeping. Galileo (1564–1642) first studied the timekeeping properties of the pendulum. He also noticed that the pendulum's period depends on its length, not mass. Galileo also used free-swinging pendulums as timers in scientific experiments to keep time for music. It turned out that Galileo's discovery is valid only for small displacements of the pendulum; in this case, the approximation leads to harmonic oscillation, while the primary interaction between pallets and gear teeth has been conserved. Since the pendulum is damped, its energy loss should be compensated by this interaction. Feedback control was necessary to establish the appropriate energy transfer phase (the phase is essential, just like when a child is pushed on a swing).

Based on Galielo's discovery, the Dutch mathematician, astronomer, physicist, and horologist Christiaan Huygens patented the pendulum clock in 1657 (The States-General of the United Provinces of the Netherlands granted patents from 1589). It is known that Huygens made a contract with clockmaker Salomon Coster, who built the clock. The accuracy of clocks increased enormously, from about 15 minutes to 15 seconds per day. The "verge and foliot" clocks were substituted with pendulums [10].

An Interplay: Synchronization of clocks. Huygens observed the synchronization of two pendulum clocks "... It is quite worth noting that when we suspended two clocks so constructed from two hooks imbedded in the same wooden beam, the motions of each pendulum in opposite swings were so much in agreement that they never receded the least bit from each other, and the sound of each was always heard simultaneously. Further, if some interference disturbed this agreement, it reestablished itself quickly. For a long time, I was amazed at this unexpected result, but after a careful examination, finally found that the cause of this is due to the motion of the beam, even though this is hardly perceptible [11]." Huygens called the phenomenon the *sympathy of the two clocks*. The clocks synchronize by transferring energy via mechanical vibrations through the coupling bar. When the vibrations exerted by one pendulum clock on the coupling bar are exactly canceled by the vibrations exerted by the other, the clocks begin to move precisely out of phase. The system consists of independent oscillators and a coupling mechanism connecting the parts and controlling their motion. In the Huygens example, a wooden beam connects the clocks.

The verge and foliot escapement has been substituted by "anchor escapement" by William Clement and Robert Hooke (1635–1703), and later an improved version (deadbeat escapement) by George Graham (1673–1751). A pendulum-driven escape engaged and released gear teeth in the same plane, reducing the oscillation amplitude and improving accuracy ($\sim$ 10sec per day). Anchor escapement established, what is called *recoil*. While the escape wheel mainly turned in one direction after impact with the lever, the escape wheel pushed it backward (recoil). Graham eliminated this recoil with his deadbeat escapement. Modern control technology uses deadbeat control to stabilize the system without overshooting.

What is essential from the book's perspective is that feedback plays a central role in both the verge-and-foliot and the pendulum-based clocks. The interaction of two subsystems in the verge-and-foliot mechanism provides a periodic motion. The first subsystem would exhibit a damped motion without any interaction. However, the second subsystem provides a constant torque input. The interaction through collision in this closed-loop system determines the period of the oscillation.

In the pendulum-based clock, the pendulum itself is damped and loses energy during each swing. The second subsystem, the escape wheel, increases the kinetic energy (similar to when somebody pushes a child on a swing).

John Harrison (1693–1776) was the "most brilliant horologist of all time". He adopted grasshopper escapement (the term was given after the discovery and characterized the jump of the lever). The interaction between the wheel and the lever has minimal friction. Harrison, an autodidact, spent decades building several clocks in a competition to determine longitude at sea, a severe marine navigation problem.

Since the local times change in the East-West direction, the knowledge of local times in two different points could be used to calculate the longitude distance between them. Sailors wanted to figure out how to navigate more precisely. The Royal Observatory in Greenwich was built in 1675 as a reference point. Harrison spent his whole life building a series of portable clocks with highly precise regulators. Some of his inventions (such as the bimetallic strips to compensate for the effects of temperature changes and the caged-roller bearing to reduce friction are still used today). He proved that a watch could measure longitude through these and other inventions.

From our perspective, the exciting thing is that the clock is the most fundamental device contributing to the modern world, and the mechanistic worldview is a **cybernetic device**. While it measures the continuous irreversible passing time, it is based on periodic motion and control processes. The mechanical clock contains a component that generates a time unit by a controlled repetitive process, and a counter keeps track of the time increments.

2.2.2.1 From the Clock to Clockwork Universe

Time is measured by the mechanical clock, which organizes the rhythm of individual and social life. Mechanical clocks were intended for signaling and were probably used in monasteries to alarm monks to pray. To maintain this regularity and order, they called the bell seven times daily to instruct the monks to pray. As monasteries and the villages surrounding them increased in number, so did the influence of bell-ringing on human actions. Lewis Mumford (1895–1990), the American historian of technology and science, said that the clock (and not the steam engine) was "the key machine of the modern industrial age" [12].

Hours and minutes started to control human activities. Benjamin Franklin (1706–1790) is alleged to have said: "Remember that time is money."

The clock soon became a model of the solar system, too. It became the symbol of regularity and predictability, i.e., of the properties of dynamically simple systems.

Nicholas Oresmus (Nicole d'Oresme) (1323–1382) saw the universe as a big mechanical clock created and set by God. While the clockwork universe received its final form in Newtonian mechanics, the mechanical clock became a symbol even earlier. The clockwork analogy was used by René Descartes (1596–1650) to describe the human body (excluding mental activity), so he was one of the founders of monistic mechanistic biology (i.e., animal bodies are machines) and the dualistic brain-mind theory. In the latter, he tried to answer the question, if the mind does not have a physical extension and matter does have one, how do they interact? Going back to clocks, Darwinian evolution theory sees natural selection as a Blind Watchmaker, as was used as a working fluid book indicates [13].

2.3 Steam Engine, Governor and the Prehistory of Cybernetics

2.3.1 Centrifugal Governor

Huygens's centrifugal governors are the links between mechanical clocks, the device, and the symbol of the cyclic worldview, and steam engines, whose operation can be explained by thermodynamics, the theory behind the irreversible macroscopic processes. The centrifugal governor was invented by Christian Huygens and used in wind power devices. They are windmills, and the governor's task was to regulate the distance and pressure between millstones.

In addition to water and wind as energy sources, steam as a working fluid was used. A steam engine is a heat engine that performs mechanical work using steam as its fluid. In a steam engine, hot steam is supplied by a boiler, expands under pressure, and part of its heat energy is converted into mechanical work. The remainder of the heat may be permitted to escape.

Thomas Newcomen (1664–1729) invented a method where a piston separates the condensing steam from the water. Early steam engines constructed for industrial applications were low-efficiency. James Watt studied Newcomen's steam engine and concluded that 80% of the steam used was wasted instead of providing motive force. Watt added a separate condenser to avoid heating and cooling the cylinder with each stroke. His invention quadrupled the potency of earlier designs.

Engines used in mills and mines needed to keep their machines turning at a constant speed. The problem was that the speed of an engine increased as its load decreased. The operator had to manually control (open or close) the steam valve whenever the water in the mine shaft changed. Watt began to think about implementing automatic and not manual control. The centrifugal (also called flyball) governor, based on the idea of a feedback control loop, solved the problem.

The evolution of the steam engine was a technological miracle that led to a new system of goods production. Factories using large machines emerged, resulting in increased efficiency, reduced time, and enormous power output. Steam engines were soon recognized as useful in transportation, resulting in the emergence of steamboats and railways. The Industrial Revolution arrived with incredible velocity. It was the most profound revolution in human history because of its massive impact on people's daily lives and the organization of societies.

2.3.2 Maxwell: Governors, Feedback Control, and Stability Studies

James Clerk Maxwell (1831–1879) (the founder of the classical theory of electromagnetic radiation) gave a mathematical analysis of Watt's flyball governor. Otto Mayr explained the importance of the paper in his "Maxwell and the Origins of Cybernetics" [14].

Two equations of motion (i.e., two second-order differential equations leading to a third-order one) were set and combined to describe the dynamic behavior of the control system. Maxwell adopted a technique (later called **linear stability analysis**) to calculate the stability conditions. He studied the effect of the system parameters and showed that the system is stable if the roots of the characteristic equation have negative real parts. This generalized method became known as the Routh-Hurwitz criterion for the local stability of fixed points. Maxwell was just one of the scientists of Queen Victoria's age who became interested in solving the problem, which later led to the birth of the modern control theory.

Victorian physicists and speed regulation: An encounter between science and technology

"The technology of speed regulation in the nineteenth century, from an economic or industrial point of view, a minor specialty of mechanical engineering, has recently received increasing attention, for conceptually it contains the origins of modern control engineering. This episode in the history of technology is remarkable for a social phenomenon: among its participants, we find not only countless engineers whose names are now forgotten but also illustrious physical scientists like Airy, Foucault, Kelvin, Maxwell, and Gibbs, men whom one would have expected to incline toward the 'pure' side of science" [15].

It is difficult to overestimate the governor's effect on technological innovation and social changes, as its role in building steam engine-based factories and transportation as steamboats and trains appeared.

While steam engines were not appropriate in the next stage of transportation (i.e., aviation), feedback control played a pivotal role in developing devices for manned flight.

2.4 Control and Stability: How did Aviation Become Possible?

2.4.1 From Birds to Airplanes

The foundation of aviation lies in the *understanding of aerodynamics*, which is the study of how air interacts with objects in motion. Pioneers like Sir George Cayley (1773–1857) and the Wright brothers, Orville Wright (1871–1948) and Wilbur Wright (1867–1912), were American aviation pioneers. They conducted extensive research on aerodynamics, including lift, drag, and airflow, which allowed them to design effective flying machines.

Bernstein sees Octave Chanute (1832–1910) as a prophetic visionary:

> "If a flying machine were only required to sail at one unvarying angle of incidence in calm air, the problem would be much easier to solve. The center of gravity would be adjusted to coincide with the center of pressure at the particular angle of flight desired. The speed would be kept as regular as possible, but the flying machine, like the bird, must rise and fall and encounter whirls, eddies, and gusts from the wind. The bird meets these by constantly changing his center of gravity; he is an acrobat and balances himself by instinct, but the problem is very much more difficult for an inanimate machine, and it requires an equipoise—automatic if possible—which shall be more stable than that of the bird."

Chanute was the Wright brothers' most important friend and correspondent in the aeronautical community.

In a paper *What are the similarities between aircraft and birds* [16], he states that we should understand that learning by imitation is a powerful mental strategy. Therefore, it is no surprise that from early on in history, humans had the dream of flying through the skies like birds.

The *shape and design of wings* play a significant role in generating lift. Engineers experimented with various wing shapes and airfoil designs to optimize lift and minimize drag. Understanding how air flows over and under wings led to efficient wing designs. *Control surfaces* like ailerons, elevators, and rudders enable pilots to control the aircraft's attitude and direction. The Wright brothers' development of a three-axis control system, which allowed for pitch, roll, and yaw control, was an aviation breakthrough.

Achieving stability in flight is critical to ensuring that an aircraft can maintain a controlled and predictable path. Engineers developed control methods to achieve *longitudinal stability* (stability along the aircraft's length) and *lateral stability* (stability from side to side), ensuring that aircraft could fly safely and predictably. The movable surface for lateral control is the famous *aileron* or *little wing* [17].

It is difficult to overestimate the *social effects of aviation* as it dramatically changed transportation and warfare. The development of powered flight marked a significant advancement in human mobility and military strategy.

Aviation revolutionized transportation by enabling faster and more efficient travel. Airplanes allowed people and goods to be transported over *long distances* in a fraction of the time it took using traditional means, such as ships or trains. Distant locations became more accessible, significantly reducing the time required for travel and increasing global connectivity. The advent of *commercial aviation* led to the airline industry's growth, making air travel a standard mode of long-distance transportation for passengers and cargo. Aviation facilitated globalization by connecting people, businesses, and cultures worldwide. It has contributed to the growth of international trade and tourism.

Aviation allowed militaries to conduct aerial reconnaissance, collecting valuable information about enemy positions and movements. More directly, using aircraft for strategic bombing allowed military forces to target enemy infrastructure, cities, and industrial centers. This changed the nature of warfare and led to the development of air raid defenses.

Gyroscopic stabilization feedback is a control system concept that utilizes gyroscopes to provide stability and maintain orientation. Gyroscopes are devices that measure and maintain rotational motion. More precisely, it measures the rate of orientation change and provides feedback to a control system, allowing it to make real-time adjustments to maintain stability. Control signals are based on the difference between the measured rotational rates and the desired orientation. By a feedback mechanism, these control signals are sent to actuators to make adjustments.

2.5   Age of Electronics: The Positive-Feedback Amplifier and the Negative-Feedback Amplifier

The positive-feedback amplifier and the negative-feedback amplifier are two different electronic amplifiers that serve distinct purposes in electronic circuits.

A *positive-feedback amplifier*, also known as a regenerative or regenerative feedback amplifier, is an amplifier that uses positive feedback to increase the gain of a circuit. In positive feedback, the output signal is fed back to the input with the same phase (or a phase shift of less than 180 degrees). Positive feedback amplifiers are typically not used for linear amplification but rather for applications in which oscillations or hysteresis are desired. They are commonly found in oscillators and signal generators. One of the most well-known positive feedback amplifiers is the Schmitt trigger [18], used in digital logic circuits to convert noisy or analog signals into clean digital signals.

A *negative-feedback amplifier*, also known as a degenerative feedback amplifier, is an amplifier that uses negative feedback to stabilize and control the gain of the circuit. In negative feedback, a portion of the output signal is fed back to the input with a phase shift of 180 degrees, which tends to reduce the overall gain. Negative-feedback amplifiers are widely used in various electronic applications, including audio amplifiers, operational amplifiers (op-amps), and most linear amplification circuits. They offer improved linearity, reduced distortion, and greater bandwidth than amplifiers without feedback. A negative feedback mechanism adjusts the amplifier's gain to maintain a stable and accurate output in changing input signals or component variations. It helped establish long-distance telephone communication; see [19] for more technical details.

2.6 Challenges of Feedback Control in Technology

Feedback control plays a crucial role in technology, providing a mechanism to regulate and maintain desired system behaviors, but it has its challenges in the space age [20].

Due to the *complexity* of technological systems, it is challenging to design effective feedback control systems since understanding dynamic behavior may be a very tough task. In addition, due to *uncertainties*, the accurate modeling of the system is a challenge. Another difficulty is the noise of the sensor system. Classical control theory assumes linearity; *nonlinearities* in the system dynamics complicate the design of feedback controllers. Additionally, achieving *stability* in a feedback control system is crucial, but it often involves trade-offs with performance metrics, such as speed and accuracy. Identifying the optimal balance between stability and performance is challenging in control system design. Incorporating *human factors* into the feedback control loop presents challenges in systems where humans interact with technology because human error is far more likely than equipment failure. Therefore, developing an optimized interface between the processes and equipment being controlled and the operators is vital to ensure safety and avoid adverse impacts on people or the technological process.

We finish the chapter by mentioning a negative example of positive feedback:

An example of dangerous positive feedback in chemical systems.

A manifestation of the dangerous positive feedback in a chemical plant environment is the temperature runaway leading to a thermal explosion.

A thermal runaway is characterized by a progressive increase in the rate of heat generation, temperature, and pressure (the latter caused by components in the reaction mass vapourising and/or decomposing to yield gaseous products at the elevated temperatures involved). Thermal runaway begins when the heat generated by a reaction exceeds the heat removal capabilities of the hardware in which the reaction is being carried out. At first, the accumulated heat produces a gradual temperature rise in the reaction mass, which causes an increase in the reaction rate. This self-accelerating process (positive feedback) may finally lead to an explosion. The problem is that an increase in temperature has a linear effect on the rate of heat transfer but has an exponential impact on the rate of reaction and subsequently on the rate of heat generation [21]. A real positive feedback.

This is just what happened in Bhopal in December 1984.

In Bhopal, much water was pumped erroneously to a methyl isocyanate tank (MIC), initiating a highly exothermic hydrolysis reaction. That resulted in a thermal runaway inside the storage tank of MIC at the Bhopal (India)Union Carbide plant in December 1984 [22]. More than 40 tons of methyl isocyanate gas leaked from the pesticide plant, immediately killing at least 3,800 people and causing significant morbidity and premature death for many thousands more. The Bhopal incident is by no means the only example of a disastrous thermal runaway occurring in a storage tank. The Seveso accident in Italy in July 1976, in which large quantities of toxic dioxin were released into the environment, occurred under storage tank conditions [23], while more or less minor runaways and explosions due to unforeseen reactions in storage tanks and vessels are relatively common.

In the period of 1962–1987, 189 industrial incidents involving thermal-runaway chemical reactions were reported to the Health and Safety Executive (Barton and Nolan, 1989). (Courtesy from Ferenc Tátrai).

2.7   Lessons Learned and Looking Forward

Control theory has been adopted implicitly and explicitly throughout the history of technological development. Negative feedback stabilizes, while positive feedback amplifies initially minor differences. Outriggers, mechanical clocks, steam engines, aviation technologies, and electronics are the main stages of progress toward the Space Travel Age. Feedback control systems played a crucial role in space exploration. The Apollo program, for example, relied on advanced control systems to navigate and stabilize spacecraft during critical phases of the mission, such as lunar landings. We also live now in the age of quantum feedback control (for a tutorial on the latter topics, see [24]). Advancements in sensors, actuators, and computing power continue to push the boundaries of what is achievable through feedback control.

From the perspective of this book, feedback control ensures the required operation. Technological systems adopt human-made control systems. The boat will not sink, the clock shows the proper time, the steam engine transfers usable energy, and the airplane flies rapidly from one continent to another. Control methods help things work as they should! Some strategy was learned from animals. As we all know, airplanes do not fly like birds, but birds still helped inspire the airplane's wing-like structure.

There are several areas in which feedback control may provide new trends. Feedback control will be vital in advancing *automation and robotics*. Intelligent feedback systems enable robots to adapt to dynamic environments and learn from experience. In biomedical engineering, feedback control systems will continue to play a critical role in drug delivery systems, wearable devices (such as smartwatches), and medical implants (from cochlear to hip implants). These systems can adapt to individual patient needs and provide personalized medical treatment. Feedback control will be fundamental in developing *autonomous vehicles*. Precise control mechanisms will be required to ensure the safety and efficiency of self-driving cars, drones, and other autonomous systems. Feedback control systems will enhance the interaction between humans and machines. Two significant areas of applications are virtual reality and brain-machine interfaces. Virtual reality is a computer-generated environment with scenes and objects that appear real, immersing the user in their surroundings. Brain-machine interfaces direct communication pathways between the brain's electrical activity and an external device, most commonly a computer or robotic limb (to learn how to play ping-pong with your brain, see, e.g., [25]).

Beyond technology, feedback control is everywhere, from biological to economic, socio-ecological, and other social systems. Next, we will turn our attention to biological control systems.

References

1. Bernstein DS (2022): Feedback control: an invisible thread in the history of technology.IEEE Control Syst Mag 22(2):2002
2. Why is the Autopilot Called "George"? (Two Prevailing Theories) https://airplaneacademy. com/why-is-the-autopilot-called-george-two-prevailing-theories/
3. Mayr O (1970) The origins of feedback control. MIT Press, Cambridge, MA
4. Mayr O (1970) The origins of feedback control. Sci Am 223(4):110–118
5. Abramovitch DL The Outrigger: A Prehistoric Feedback Mechanism. In: IEEE control systems
6. https://www.youtube.com/watch?v=cPAbx5kgCJo+
7. Needham J (1959) The missing link in horological history: A Chinese contribution. In: Proceedings of the royal society of London. Series A, vol 250, pp 147–179
8. Needham J, Wang L, De Solla Price D (1960) The great astronomical clocks of Medieval China. Cambridge Univ. Press, Heavenly Clockwork
9. Landes D (1983) Revolution in Time: Clocks and the making of the modern world. Belknap Press of Harvard Univ. Press, Cambridge, MA
10. https://www.historyofinformation.com/detail.php?id=3068
11. Pikovsky A, Rosenblum M, Synchronization. http://www.scholarpedia.org/article/ Synchronization
12. Mumford L (1934) Technics and civilization. Harcourt, Brace and Company, New York
13. The Blind Watchmaker: Why the Evidence of Evolution Reveals a Universe without Design
14. Mayr O (1971) Maxwell and the origins of cybernetics. Isis 62(4):425–444
15. Mayr O (1971) Victorian physicists and speed regulation: An encounter between science and technology. The Roy. Soc. 26(205–228):1971
16. What are the similarities between aircraft and birds https://www.grupooneair.com/similarities-between-aircraft-and-birds/#the_wings_an_essential_element
17. Abzug MK, Larrabee EE (2009) Airplane stability and control. Cambridge University Press, A history of the technologies that made aviation possible

18. Schmitt OH (1938) A thermionic trigger. J Sci Instr 15(15):24–26
19. https://electronicscoach.com/feedback-amplifier.html
20. Bernstein D (2022) Facing future challenges in feedback control of aerospace systems through scientific experimentation. https://arc.aiaa.org/doi/full/10.2514/1.G006785
21. Barton IA, Nolan PF (1989) Incidents in the chemical industry due to thermal runaway reactions. IChemE Symposium Series No. 115:1989
22. Barton N (1989) Incidents in the chemical industry due to thermal runaway reactions. Barton, I A, Nolan, P F. IChemE Symposium Series No. 115:1989
23. Theophanous TG (1983) Theofanous, the physicochemical origins of the Seveso accident–II: Induction period. Chem Eng Sci 38(10):1631–1636
24. Altafini C, Ticozzi F (2012) Modeling and control of quantum systems: An introduction. IEEE Trans Autom Control 57(8):1898–1917. https://doi.org/10.1109/TAC.2012.2195830
25. Nayvelt L (2019) Playing Ping-Pong with my brain. https://medium.com/@nayvelt.lina/playing-ping-pong-with-my-brain-49025f044b9f

Chapter 3
Feedback Control in Biological Systems

Abstract Feedback control is a fundamental tool at every level of the biological hierarchy, from cellular to socio-ecological systems. It ensures homeostasis by adopting a general mechanism for restoring certain states after a small perturbation. Dynamical diseases occur due to the impairment of control systems. The theory of nonlinear dynamics offers a mathematical framework to analyze pathological temporal patterns. It aims to find control strategies to shift the physiological parameters back into normal ranges.

3.1 Feedback Control in Biology: An Overview

In Chap. 2, we demonstrated that feedback control was used in different places for thousands of years to revolutionize everything from transportation via measuring time to establishing modern industrial societies. It turned out that evolution generated numerous control mechanisms at very different levels of hierarchical organization, from molecules to ecological systems.

Claude Bernard (1813–1878) and Walter Bradford Cannon (1871–1945) recognized that our internal physiological state shows remarkable stability and provides the biological basis for our ability to live an independent life. The concept of **homeostasis** was born, and we start this chapter by discussing this fundamental notion.

The following section gives brief biological feedback control case studies. It is remarkable how many cellular mechanisms adopt feedback control; we will mention two. The celebrated *lac operon* model of Francois Jacob (1920–2013) and Jacques Lucien Monod (1910–1976) discovered the first genetic regulatory mechanism, and the work led the authors to be awarded the Nobel Prize in Physiology in 1965. The discoveries of molecular mechanisms controlling our biological clock related to the *circadian rhythm* by Jeffrey C. Hall (1945), Michael Rosbash (1944), and Michael W. Young (1949) also resulted in a Nobel prize (2017) for the discoverers. As living organisms show periodic behaviors with very different frequencies, it is exciting to see how the interaction of different oscillators leads to synchronization and other dynamic phenomena.

© The Author(s), under exclusive license to Springer Nature Switzerland AG 2024
P. Érdi, *Feedback*,
https://doi.org/10.1007/978-3-031-62439-1_3

Three more illustrative examples will follow: (1) thermo-regulation is pivotal to mammalian life as a homeostatic mechanism that maintains internal temperature within a narrow range; (2) the pancreas maintains glucose homeostasis within a minimal 4–6 mm interval; (3) Cannon also described the fight-or-flight response that occurs in response to a perceived harmful event, attack, or threat to survival.

In the last section, dynamical diseases will be studied. Mathematical physiology develops models: (i) to uncover pathological mechanisms; (ii) to identify the dynamic changes in those physiological variables that can be easily monitored to warn of the emergence of illness; (iii) to develop therapeutic control strategies.

3.2 Homeostasis: A General Concept

3.2.1 A Capsule Pre-History of Hoemeostasis

Very often, when we start to analyze a modern concept, we first turn to search for its roots in ancient Greek times. Alcmaeon of Croton stated that the *equality* (isonomia) of the powers (wet, dry, cold, hot, bitter, sweet, etc.) maintains health but that *monarchy* among them produces disease [1]. Using modern language, he adopted a principle of *balance of opposites* to explain health and disease. If you wish, Alcmaeon was an interdisciplinary scientist who saw an analogy between sick people and sick society: inequality of power leads to tyranny in a political system and disease in the body. Greek medical writers generally accept that health depends on a balance of opposing factors in the body. Alcmaeon wrote about a broad range of topics in physiology, such as embryo development, sleep, and death.

To continue the short pre-history of homeostasis [2], we turn now to Hippocrates, who stated that the balance among *humors* (i.e., chemical systems regulating human behavior) is fundamentally essential to preserve health. Hippocrates mentioned four humors: blood, phlegm, yellow bile, and black bile. Pain emerges when one of the substances appears in "too much" or "too little" quantity. Since Hippocrates knew about balance and imbalance, it is reasonable to assume that Hippocrates knew at least implicitly that a healthy life is based on some stabilizing mechanism. He and the school of medicine that followed him can be considered the originators of the notion of "homeostasis". Galenus (Galen, 129–216) championed medical research in the Greco-Roman area, and his views dominated Western medical science for more than 1,300 years. Galen applied the four humors theory to classify people's character: an imbalance of each humor corresponded with a particular human temperament (blood—sanguine, black bile—melancholic, yellow bile—choleric, and phlegm—phlegmatic). Galen believed blood was not conserved but constantly consumed and restored.

"Anatomy is to physiology as geography is to history; it describes the theatre of events." This quote is derived from Jean Francois Fernel (ca. 1497–1558). He made comparative studies of healthy and pathological bodies and stated that health is maintained as the parts of the body work together. William Harvey (1578–1657) realized that, as opposed to Galen's thesis, "the blood in the animal body moves around in a circle continuously and that the action or function of the heart is to accomplish this by pumping" [3]. Harvey worked during and soon after modern science was born by Galileo and Kepler, and observations have been supplemented by experiments and computations [4]. As Copernicus dropped the Ptolemy model, which dominated for about 1500 years, Harvey somewhat similarly rejected Galan's ideas about the emergence and disappearance of blood. Two critical scientific paradigms emerged.

3.2.2 Physiological Homeostasis

The precondition of continuing the healthy functioning of our body is the existence of corrective mechanisms. Such mechanisms help us to live and adapt ourselves to changing environments. Claude Bernard (1813-1878), a French physiologist, made significant contributions to establishing modern experimental, methodological, and conceptual bases of contemporary physiology. He introduced the notion of regulating the internal environment (often mentioned as *milieu interieur* to give credit for the French origin). The stability of the internal environment is a necessary condition for independent life. Bernard recognized that the body executes mechanisms that operate in an organized fashion to maintain a relatively constant temperature and blood glucose concentration. Internal stability (i.e., the ability to show resistance) is vital for the organism's health [5, 6]. The living body needs a surrounding environment but somehow preserves its relative independence. This independence derives from the fact that in a living being, tissues are withdrawn from direct external influences and protected by an internal environment, which is constituted mainly by fluids circulating in the body.

Walter Bradford Cannon (1871–1945) developed the concept of homeostasis from the earlier idea of Claude Bernard of the internal environment and popularized it in his book *The Wisdom of the Body* [7]. Cannon made fundamental statements about the general features of homeostasis, as follows:

---

(1) Constancy in an open system requires mechanisms to maintain this constancy. Cannon based this proposition on insights into how steady states such as glucose concentrations, body temperature, and acid-base balance were regulated. (2) Steady-state conditions require that any tendency toward change automatically meets with factors to resist that change. An increase in blood sugar results in thirst as the body attempts to dilute the sugar concentration in the extracellular fluid. (3) The regulating system that determines the homeostatic state consists of several cooperating mechanisms acting simultaneously or successively. Insulin, glucagon, and other hormones regulate blood sugar, controlling its release from the liver or tissue uptake. (4) Homeostasis does not occur by chance but results from organized self-government [8].

3.2.2.1   Stability, Thermodynamics, and Homeostasis

There is a general mechanism for restoring certain states after a small perturbation. These states are called *stable* equilibria, and while their definition comes from physics and chemistry, the concept can also be applied to social systems (also discussed in our book **Repair** [9]).

The first and second laws of thermodynamics reflect *constancy* and *change*. The first law states that energy is conserved (neither created nor destroyed but is transferred and transformed). The second law states that macroscopic processes are irreversible and have a direction. At that time, it was not trivial that the laws of physics could be applied to living systems. Such systems often involve energy transfer. Consequently, the first law of thermodynamics is directly relevant to homeostasis.

The role of food: Humans and other living systems convert food to the energy needed to maintain normal body functions. Human metabolism involves converting forms of energy in food into other forms needed to live. Metabolism converts food's "chemical" energy to work and heat energy, which is needed to maintain body temperature and perform physical activities. Food is a basic building block that supplies living things with the energy they convert for homeostatic processes. Homeostatic functions such as calorie intake, sleep, and temperature regulation all operate within the bounds of the first law of thermodynamics.

The second law of thermodynamics, i.e., the law of irreversibility, can also be identified. While body fat can be converted to do work and produce heat transfer, work done on the body and heat transferred into it cannot be converted to body fat. Otherwise, instead of having lunch, we may sunbathe (the combination of the two sounds fantastic) or walk downstairs.

Claude Bernard employed thermodynamics in medicine and physiology with his concept of feedback regulation of the internal milieu in living organisms. He realized that external variations are compensated for and equilibrated at each instant to maintain stability.

In 1884, Henry Louis Le Chatelier (1850–1936) proposed an axiomatic concept of thermodynamic equilibrium (now known as Le Chatelier's Principle) to describe chemical systems in equilibrium. Le Chatelier's principle is based on observing the chemical equilibria of reactions and states that changes in physicochemical conditions—such as temperature, pressure, volume, or concentration of a system—will result in opposing modifications within the system to achieve a new equilibrium state. A textbook example of Le Chatelier's principle is the Haber process for producing ammonia. The pressure increase favours the forward reaction and increases the yield of ammonia.

Walter Cannon's concept of homeostasis is a central tenet of physiology: the inherently unstable properties of a living system are regulated such that they remain relatively constant in a state of dynamic equilibrium. Our bodies are intrinsically unstable, subjected to externally and internally induced injuries that require homeostatic control via repair processes.

Dyshomeostasis is a state of imbalance or disrupted homeostasis in which normal physiological processes are disturbed. Specifically, during aging, among others, sleep patterns change. Dyshomeostasis refers to alterations in regulating and balancing sleep-wake cycles and associated neural circuits.

Living organisms require a continuous input of energy for their maintenance. Life is based on metabolic processes that extract energy from the environment and make it available to support life. This energy metabolism is based on and regulated by underlying thermodynamics. Many details of the kinetic mechanism and thermodynamic basis of homeostasis are known, and further details can be found in [11].

3.2.3 Homeostasis Most Likely Does Not Exist on a Global Level: The Gaia Metaphor

There is a theory, model, or, more likely, a metaphor that states that homeostasis exists at the level of the whole Earth. Here is the short article Encyclopedia Britannica writes:

Gaia hypothesis, model of the Earth in which its living and nonliving parts are viewed as a complex interacting system that can be thought of as a single organism. The Gaia hypothesis is named after the Greek Earth goddess, developed in 1972 by British chemist James E. Lovelock and U.S. biologist Lynn Margulis. It postulates that all living things have a regulatory effect on the Earth's environment that promotes life overall; the Earth is homeostatic in support of life-sustaining conditions. The theory is highly controversial.

The debates around the Gaia theory/hypothesis [10] (named after the ancient Greek goddess of Earth) were highly publicized. It differs significantly from the usual conflict between pseudoscientists and the scientific community. James Lovelock (1919–2022) was a British scientist who created and developed many scientific instruments and became a Fellow of the Royal Society in 1974. Lynn Margulis (1938–2011) was an American evolutionary biologist selected to be a member of the US National Academy of Sciences in 1983. President Bill Clinton awarded her with the National Medal of Science in 1999.

The Gaia hypothesis suggests that the Earth (composed of living organisms and an inorganic environment) is a self-regulating system, showing the features of living systems. The statement that *Earth is alive* comes from a combination of observations and wishful thinking. The hypothesis suggests that the concept of homeostasis can be extended from the individual level to the whole biosphere and that the Earth should be considered a super-organism.

Lovelock knew a hypothesis was more convincing if an underlying mechanism supported it. Therefore, he offered a causal mechanism, a mathematical model studied by computational simulations to demonstrate that feedback mechanisms can evolve from the actions or activities of self-interested organisms. The simulation model, Daisyworld [12], can be considered a toy model. It defines a hypothetical world orbiting a star whose radiant energy slowly increases or decreases. It is meant to mimic essential elements of the Earth-Sun system. Two varieties of daisies were assumed to be life forms: black and white. White-petaled daisies reflect light, while black-petaled daisies absorb light. The simulations show how the two daisy populations and the surface temperature of Daisyworld change as the sun's rays show substantial growth. The surface temperature of Daisyworld remained almost constant over a broad range of solar output. To use a different language, the model supports a self-regulating mechanism to provide viable conditions. In an extreme form, the Earth is assumed to be a superorganism: a holistic feedback system in the biosphere. Life forms regulate temperature and proportions of atmospheric gases to life's advantage.

The main statements of the hypothesis in a simplified form: (i) The conditions of the Earth are highly favorable to creating and maintaining life; (ii) the chemical composition of the atmosphere and the sea changed due to life; (still) (iii) The Earth's environment remained stationary over geological times.

There were many pros and cons to accepting and rejecting the hypothesis. The situation is far more complicated, but by and large, I agree with the verdict of Toby Tyrell, the British Earth system scientist at the National Oceanography Centre at the University of Southampton [13, 14]. He states that the hypothesis is beautiful but flawed. The records on the past ice ages challenge the first statement. Terrestrial vegetation was reduced to about half that of the warmer interglacial periods. The second claim might be valid, but there is no apparent reason to believe that the changes increase the chance of Earth's habitability. Records on climate cycles do not support the third claim. Most likely, there is no homeostatic mechanism at the global level. Our Earth is not as robust as it should be. Our best option is to accept that our Earth and its environment are fragile and that we must defend it.

3.3   Applications: From Cellular Biology to Socio-ecological Systems

3.3.1   Cellular Biology

3.3.1.1   Feedback Regulation in the Lactose Operon: A French Success Story

The Nobel Prize in Physiology or Medicine 1965 was awarded jointly to three French scientists François Jacob, André Lwoff, and Jacques Monod "for their discoveries concerning genetic *control* (my *italic; P.E*) of enzyme and virus synthesis". Francois Jacob and Jacques Monod figured out how bacteria controlled the production of an enzyme called beta-galactosidase. This feedback and the negative regulation system became the *lac operon* and was the first model for controlling protein production. Cell biologists use their terminology to describe positive and negative feedback control. They adopted such expressions as inducer, promoter, repressor, silencer, activator etc..

An *inducer* is a molecule that regulates gene expression. An inducer functions in two ways, namely: (i) By disabling the so-called *repressors*. The gene is expressed because an inducer binds to the repressor. The binding of the inducer to the repressor prevents the repressor from binding to the operator. RNA polymerase (an enzyme that catalyzes the chemical reactions that synthesize RNA from a DNA template) can then begin to transcribe operon genes. (ii) By binding to activators. *Activators* generally bind poorly to a DNA sequence unless an inducer is present. The activator binds to an inducer, and the complex binds to the activation sequence and activates the target gene. Removing the inducer stops transcription. Because a small inducer molecule is required, the increased expression of the target gene is called induction. The lactose operon is one example of an inducible system [15]. Loosely speaking, promoters and repressors implement positive and negative feedback.

Jacob and Monod outlined a theory of genetic control in prokaryotes (prokaryotes are simple cells that don't contain a nucleus, while eukaryotes do) in 1961 [16, 17]. The Operon model is the classical model for cellular metabolism, growth, and differentiation (for the legacy and historical analysis of this seminal work, see [18]). There are well-accepted detailed mathematical models [19] which, by taking into account the network structure of the lactose operon regulatory system (Fig. 3.1, can reflect the fundamental bistable property of the system.

The lac operon became a paradigm for gene regulation for bacteria and all biological systems from simple phages to complex humans. Even earlier, classical experiments [20] suggested the *all-or-none* character of gene expression in response to an inducer. The lac operon in Escherichia coli has been studied extensively and is one of the earliest gene systems found to undergo both positive and negative control. It is known to exhibit bistability because the operon is induced or uninduced.

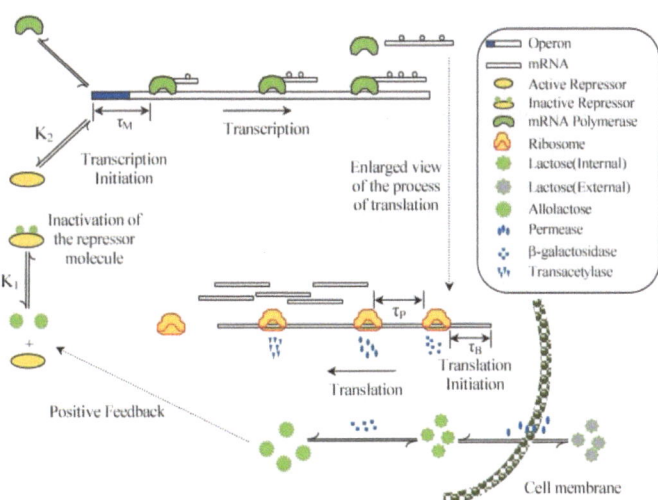

Fig. 3.1 Schematic representation of the lactose operon regulatory system [21]

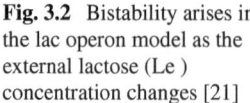

Fig. 3.2 Bistability arises in the lac operon model as the external lactose (Le) concentration changes [21]

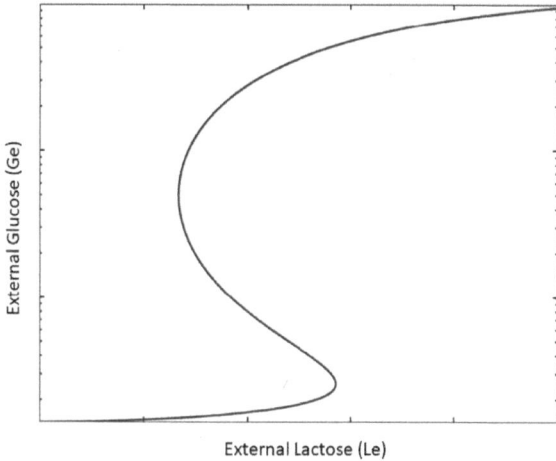

External Lactose (Le)

(Figure 3.2 shows an idealized bistability in a simple model with only one control parameter, the external lactose concentration).

More sophisticated models in the lac operon have been established [22]. Bistable regions were identified in the two-dimensional parameter space formed by the external lactose (Le) and external glucose (Ge) concentrations. The model also predicted that bistability disappears for deficient Ge concentrations.

3.3.2   An Interplay: Bistability as a Very General Phenomenon

Bistability occurs in many natural, social, and technological systems when a system can exist in either of two so-called steady states and when there is an abrupt jump from one state to another. Bistability is a property of specific nonlinear systems. Such phenomena were demonstrated and analyzed in different fields, from phase transition in physics to price jumps and the interpretation of ambiguous figures due to multistable perceptions. Abrupt jumps from one state to the other were subjects of interdisciplinary studies under the names of "phase transition", "synergetics" and "catastrophe theory".

In the 1970s, three disciplines, each interdisciplinary and emphasizing "nonlinearity," became fashionable in Europe. The implicit connotation of "nonlinearity" was that "linear" was less interesting and that nonlinearity leads to different and lovely patterns in time and space. Three European schools dominated the field. The theory of "dissipative structures" grew out of non-equilibrium thermodynamics and was named by Ilya Prigogine and his "Brussels school." Hermann Haken (Stuttgart) used the term "synergetics" to deal with systems in which order is an emergent property

Fig. 3.3 Bistability and hysteresis

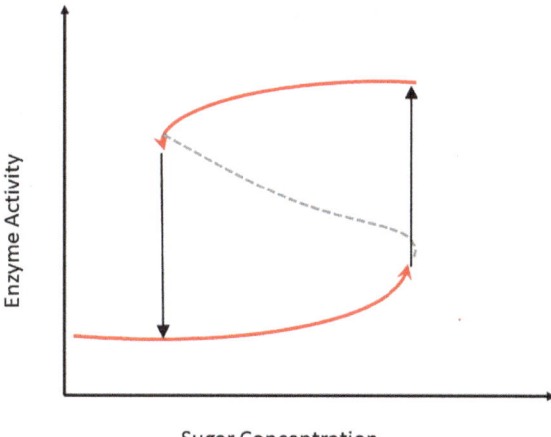

of macroscopic systems due to the interactions of elementary constituents. "Catastrophe theory" is a mathematical theory for classifying abrupt qualitative changes in the behavior of a system due to small changes in the circumstances. It emerged from the works of the French mathematician René Thom. Thom's approach was deeply deterministic, while the other two schools took into account random effects and fluctuations to get ordered structures [24].

It is often mentioned that there is a direction-dependent phenomenon, i.e., hysteresis. This refers to when the jump from the regime of "low" fixed points to the regime of "high" fixed points and the jump back from the "high" regime to the "low" regime do not happen at the same parameter values. The phenomenon should not be overemphasized since the parameters do not depend on time. It is more informative to say that a bistable system can classify the set (actually the interval) of the initial values. Figure 3.3 is an example, and we will discuss more bistable phenomena later in this book.

3.3.2.1 Biological Clocks

The rotation of our planet has encouraged life on Earth to adapt. We have long known that all living beings, including people, have inherent biological clocks that enable them to anticipate and adjust to a daily rhythm. But how does this clock function? The Nobel laureates of 2017, Michael Rosbash, Michael W. Young, and Jeffrey C. Hall gave a clear explanation. Their findings explain how plants, animals, and humans modify their biological rhythms to match the Earth's planetary revolutions. Simply put, they discovered a gene that regulates the typical daily biological rhythm using *fruit flies* as a model organism. They demonstrated that this gene produces a protein that builds up in cells at night and is destroyed during the day. Later, they discovered other protein parts of this machinery, revealing the mechanism governing the **self-sustaining clockwork** within the cell. Nearly every tissue and organ contains biological clocks. Researchers have identified similar genes in fruit flies and mice, plants, fungi, and other organisms that make the clocks' molecular components. We now understand that multicellular organisms, including humans, have biological clocks that operate according to the same principles.

Our inner clock adapts our physiology to the dramatically varied phases of the day with great precision. The clock controls essential functions such as hormone levels, body temperature, metabolism, sleep, and behavior. When there is a momentary mismatch between our exterior environment and this internal biological clock, such as when we travel across many time zones and experience "jet lag", our well-being suffers. There is also evidence that a persistent misalignment between our lifestyle and the rhythm dictated by our inner timekeeper is linked to an elevated risk of various disorders [23].

In the brain, a *master clock* coordinates all the biological clocks in a living thing, keeping the clocks in sync. In vertebrates, including humans, the master clock is a group of about 20,000 neurons that form a structure called the suprachiasmatic nucleus (SCN). The SCN is in a part of the brain called the hypothalamus and receives direct input from the eyes.

Mathematical models have been proposed to explain the behavior of the circadian oscillator. Shortly after François Jacob and Jacques Monod developed their first model of gene regulation, Brian Goodwin (1931–2009) suggested the first model of a genetic oscillator, showing that regulatory interactions among genes allowed periodic behavior to occur. Goodwin developed a straightforward model for limit cycle oscillations caused by negative feedback on gene expression. A given gene is transcribed into mRNA and then translated into a peptide. The latter acts as a repressor: it inhibits mRNA synthesis. A nonlinear, hyperbolic function was assumed to describe the decrease of repression with increasing inhibitor concentration and determine the transcription rate. Combined experimental and theoretical studies explain how the interplay between positive and negative feedback plays a role in generating and controlling circadian rhythms [25, 26]. Figure 3.4 shows examples of short positive and negative feedback loops. $X \rightarrow Y$ means X *activates* Y and $X \dashv Y$ denotes X *inhibits* Y.

Theoretical biochemists discovered and classified the conditions sufficient to generate oscillators. Specifically, *time delays* in negative feedback destabilize equilibrium points and may lead to sustained oscillatory behavior [26]. Negative feedback

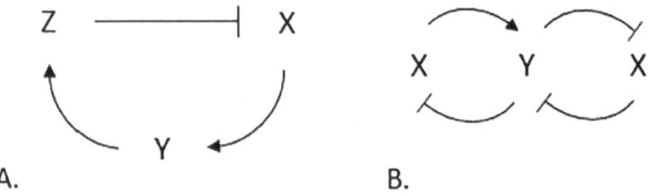

Fig. 3.4 Activating-inhibitory motifs

reduces system deviations from a steady state, while a sufficiently long delay before the feedback acts as a destabilizing factor.

From Huyghens to Kuramoto: fireflies and crickets

The coupling of oscillators may lead to intricate dynamic behaviors, such as *synchrony*. Huygens' observation about the synchrony of two mechanical clocks was mentioned earlier in 2.2.2. Collective synchronization was first studied mathematically by Norbert Wiener, who, as we remember, was the father of cybernetics. The Japanese physicist Yoshikiki Kuramoto established a celebrated model in which a set of oscillators undergo synchronization as the coupling strength increases. Each oscillator has its intrinsic natural frequency, and the effect of coupling is that the oscillators tend to increase or decrease their frequency to approach the *mean phase* of the other oscillators.

The Kuramoto model [27] was initially motivated by the phenomenon of collective synchronization: a vast number of oscillators spontaneously lock to a common frequency despite the unavoidable differences in the natural frequencies of the individual oscillators.

There are many biological examples: circadian pacemaker cells in the SCN, a network of pacemaker cells in the heart, and metabolic synchrony in many biochemical systems. There are well-studied problems of social coordination, such as synchronously flashing fireflies and crickets that chirp in unison. Flashes from the male fireflies started in one location—once there was a critical density of individuals—and then propagated through the swarm [28]. Fireflies perceive visual cues from others near and far to attract females on the ground. The big open question is whether there are leaders and followers. The same question exists in the society of crickets.

Bush crickets sing in unison, in almost (but just nearly!) perfect synchrony. Singing in a group might help keep males on a steady rhythm—a trait that female crickets preferred. The sound certainly attracts females, so cooperation is vital for male crickets. However, cooperation coexists with the competition. Males frequently timed their signals as leader or follower with an average time lag of about 70 ms. Females selected males in such choruses based on signal order and duration [29]. The timing of the male signals in *imperfect synchrony* has fundamental consequences for a male's attractiveness because females prefer leading signals over identical signals broadcast with a time lag. The problem of synchrony attracts mathematicians, and a beautiful paper discusses the improvements in the Kuramoto model [30].

3.3.3 Temperature Regulation

Thermoregulation is a textbook example of negative feedback control. Mammals (including humans) and birds maintain body temperature with fine-tuned, controlled self-regulation. Their temperature does not depend on external temperatures. Thermoregulation is a specific example of homeostasis since preserving a stable internal temperature is necessary for survival. Animals that rely on their external environment for body heat are called ectotherms. At the same time, endotherms use thermoregulation to maintain their internal body temperature even when their external environment changes. The optimal temperature at which the human body's systems function is around 37 degrees Celsius.

Thermoregulation is crucial to human life; without thermoregulation, the human body would cease to function. It also adapts to the body's response to infectious pathogens. The mechanism of thermoregulation in living systems is much more complicated than the operation of a thermostat. Biological control systems have three parts: afferent sensing, central control, and efferent responses. First, there are hot- and cold-sensing receptors throughout the whole body. Second, the hypothalamus is the central regulator. It regulates and fine-tunes a complex set of temperature-control activities. If it senses internal temperatures growing too hot or too cold, it will automatically send signals to the skin, glands, muscles, and organs. The hypothalamus helps balance body fluids and salt concentrations and controls the release of temperature-related chemicals and hormones. Third, efferent responses are the behaviors that humans can engage in to regulate their body temperature. As we all know, sweating is one mechanism the body can use to cool itself. Conversely, a shivering reflex in a cold environment results in skeletal muscles contracting and producing heat. To wear more or fewer layers of clothing is our behavioral response to temperature change.

Winter swimming in cold water supplemented with hot sauna sessions is well-known in North Europe and other cultures. Brown fat is body fat that keeps us warm when we get cold. It also stores energy and helps our body burn calories. Brown fat starts working in freezing temperatures. Brown fat activates right before we start to shiver. Brown fat also burns calories and stores energy. According to a recent study on the benefits and risks of cold water swimming [31] "When cold water swimming is practiced by experienced people with good health in a regular, graded, and adjusted mode, it appears to bring health benefits. However, there is a risk of death in unfamiliar people due to the initial neurogenic cold shock response or a progressive decrease in swimming efficiency or hypothermia."

3.3.4   Blood Sugar Concentration Regulation

A famous example of negative feedback on biological systems is the control of blood sugar. The small intestine will absorb glucose from the digested food after a meal. There is a consequent *rise* in blood glucose levels. Increased blood glucose levels stimulate the beta cells of the pancreas to *produce* insulin. Insulin triggers liver, muscle, and fat cells to absorb and store glucose. The blood glucose level thus *goes down* when glucose is absorbed. When insulin levels *fall below the limit*, there is not enough stimulus to release it, and beta cells are *shutting down*. Figure 3.5 illustrates the operation of the feedback loop.

3.3.5   Fight-or-Flight Response

When our ancient ancestors faced some terrifying danger in their environment, they had to make a rapid decision: either they could fight or flee. The fight-or-flight response evolved as a survival mechanism, enabling humans and other mammals to react quickly to life-threatening situations. The physiological mechanisms of the *fight-or-flight response* were described in the 1920s by Walter Cannon, who realized that a chain of rapidly occurring reactions inside the body helped to mobilize the body's resources. The physiological response is associated with releasing hormones that prepare our body to either deal with a threat or run away to safety [32, 33].

First, the body's sympathetic nervous system is activated due to the fast release of hormones. Second, the sympathetic nervous systems stimulate the adrenal glands, triggering the release of catecholamines (adrenaline and noradrenaline). Third, heart rate, blood pressure, and breathing rate are increased. According to standard conditions, the threat disappears after a 20-minute to 60-minute relaxation process.

How we respond to stress and danger is essential, and the fight-or-flight response helps us with that. The body prepares to fight or run away when it perceives a threat. Knowing that our response can come from real or imaginary dangers is essential. If we prepare our bodies for action, we will be more ready to react as necessary when there is pressure. The worry caused by a problem can sometimes be good because

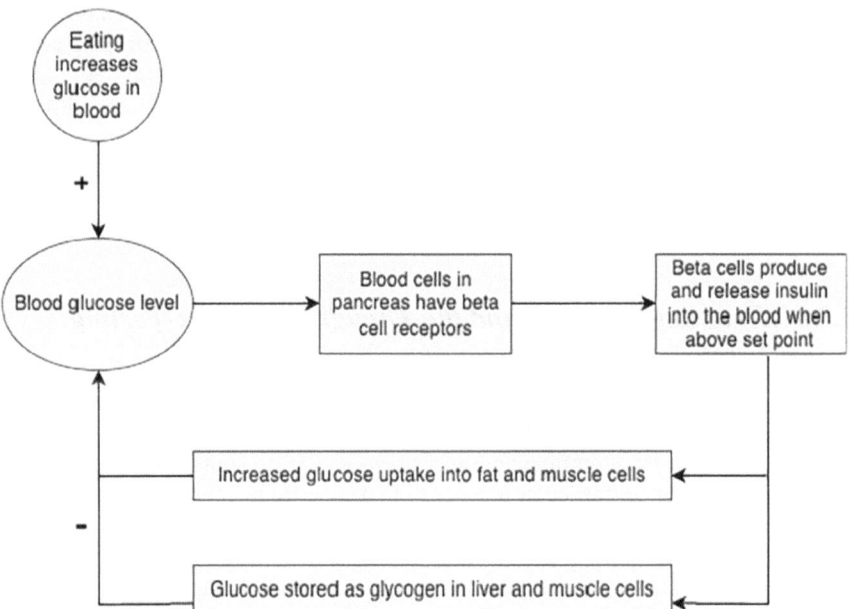

Fig. 3.5   Feedback control of blood sugar concentration. From https://www.mdpi.com/2571-5577/ 3/3/31 Fig. 2. A conceptual overview of the biological perspective on how a β-cell achieves glucose control

it can help us deal with danger better. Feeling this way can make us do better in situations we need to do well, like at work or school.

The fight-or-flight response happens without us thinking, but sometimes it is useless. We sometimes react like this even if there is no danger. If someone is afraid of heights, they might start feeling stressed out when they have to go to the top of a tall building for a meeting. Their speeding heartbeat and heavy breathing could cause unnecessary and potentially harmful side effects, like a panic attack.

One way to address difficult situations is to understand how our body naturally reacts when we feel scared or stressed. When we feel anxious, we should find ways to become calm and relaxed. Health psychologists study how stress affects our body and mind, an essential topic in their field. They want to help people overcome stress and be healthier and more productive. The fight-or-flight response is now considered part of the first stage of *general adaptation syndrome* developed by Hans Selye (1907-1982), a stress theory.

Fight, Flight, Freeze, Fawn

In addition to the originally suggested two extreme responses, other possible and infrequent reactions exist. **Freezing up** is a very typical response, too. It happens when we cannot choose immediately between fight or flight. We might experience dissociation or detachment as we cannot defend ourselves or escape the threatening situation. Sometimes, freezing allows us to find more time to fight or flee; other times, it can leave us at a greater risk of harm [34].

Another possibility is the **fawn** response. It is a secondary strategy after an unsuccessful fight, flight, or freeze attempt. A few individuals react to undermining people by turning to people-pleasing. They might endeavor to compliment or calm down the individual damaging them, maybe endeavoring to help the disposition with self-deprecating humor. Individuals who grew up with narcissistic guardians regularly drop back on the grovel reaction since it was valuable in exploring their difficult childhoods.

3.3.6   Emotional Control and the Thoughts–Actions–Feeling Circle

Feeling regulation is our capacity to apply control over our enthusiastic state. The objective is to diminish outrage and uneasiness, cover up signs of our fear or pity for others, or center on particular reasons to feel cheerful or calm. One transcendent way is the constrained reconsidering of challenging circumstances.

Thoughts are mental cognitions—our ideas, suppositions, and convictions around ourselves and the world around us. They incorporate the viewpoints we bring to any circumstance or involvement that color our point of view (for superior, more regrettable, or impartial). An illustration of a long-lived thought is an attitude created as considerations are repeated and fortified.

Whereas contemplations are formed by genetics, life experience, or education, they generally implement conscious control. In other words, knowing your thoughts and attitudes allows you to change them.

It may be valuable to think of **emotions** and **feelings** as a stream and involvement of sentiments, such as anger, fear, joy, or sadness. Whereas feelings are all-inclusive, each individual may be involved and react differently to them. A few individuals may struggle to understand what feeling they are encountering. Feelings associate us with others and offer assistance in developing solid social bonds. This may be the evolutionary purpose of emotions—people who could make stable bonds and enthusiastic ties became a part of a community and were more likely to discover the support and security vital for survival.

Behavior is characterized by what we do and how we perform **actions**. Our thoughts and feelings have a significant impact on our actions. But our actions also have consequences for our feelings and thoughts. In 1.1.2, we mentioned the concept of circular causality, one of the basic principles of cybernetics. Our thoughts, feelings, and behaviors are interconnected and determine the quality of our life. When each subsystem influences the others, as Fig. 3.6 shows, we speak about *network causality*.

As a college professor, I often receive emails from students requesting a post-ponement of their presentation by a few hours. What might be the mechanism of their action? Based on my conversation with these students, the initial driving is most likely their morning negative thoughts about their presentations. "People will not like my talk." The negative thoughts imply negative feelings: some students feel extreme anxiety, possibly even panic. These feelings induce action: "Whatever happens, I will not show up! I should write some credible excuse to the professor." Whatever is in my mind, I generally accept the excuse and write back. "Okay, you can do it on Friday." So, the student generates a new thought: "Well, I gotta two more days. Let's drink a beer!"

Fig. 3.6 The thought–emotion–action network

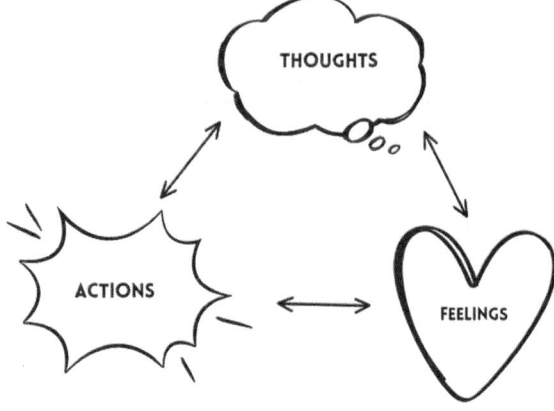

If you wish, you can see this story with a positive message. Instead of having a *vicious circle*, which intensifies her negative feelings, she managed "to solve the problem" for at least 48 hours.

People with borderline personalities have difficulties breaking the thought-feeling-action circle, and the positive feedback loop amplifies the situation with potentially dramatic consequences. Cognitive Behavior Therapy (CBT) is now widespread. Regarding its scientific reputation, according to a highly cited paper [35], "the evidence-base of CBT is robust. However, additional research is needed to examine the efficacy of CBT for randomized-controlled studies."

The key takeaway is that from a control-theoretic perspective, CBT aims to modify emotions, say, to turn sadness into happiness, or at least neutral. To make changes in our emotional state, we should make changes in our thoughts. "I am sad because I am sitting alone at home. Going to a pub without a partner is not fun, but it may be better than nothing." An action might follow a thought. "Okay, I'll go to the Deep Blue Pub. They play Celtic music Thursday nights, and there are chess tables and some hustlers."

3.4 Dynamical Diseases

3.4.1 From Normal to Pathological Dynamics and Back

As it became clear that both regular oscillatory and irregular chaotic temporal patterns occur both in nature and society and the theory of dynamical system developed, it soon became apparent that the new theoretical framework of *nonlinear science* offers a new perspective to understand and treat a large class of diseases.

In 1977, Leon Glass and Michael C. Mackey introduced the idea that certain diseases, which they labeled dynamical diseases, arise when an intact physiological control system operates in a range of control parameters that leads to abnormal dynamics and human pathology. This idea was closely associated with the mathematical concept of bifurcations occurring in dynamical systems as certain parameters change. This observation opened the door for "mathematicians at the bedside." Over the last four decades, the impact of mathematical insights on human illness has been palpable. Mathematical contributions have provided important insights into the nature of many life-threatening illnesses, including cardiac and respiratory arrhythmias, epileptic seizures, periodic hematological diseases, falls in the elderly, and certain psychiatric disorders and diseases [36].

3.4.1.1 Temporal Aspects: The Concept

The trademark of a dynamical illness/disease may be abrupt changes in the temporal pattern of physiological variables. Such changes are frequently related to changes in anatomical structures or physiological parameters. The noteworthy point of recognizing a dynamic illness is that it ought to be possible to create restorative therapeutic procedures based on our understanding of the dynamics. The hope is that control strategies can be found to shift the physiological parameters back into normal ranges.

Arrhythmias are well-known in cardiology, neurology, and respirology. A significant fraction of patients exhibit such abnormalities in the dynamics of a pathophysiological process that too often leads to death, as in cardiac arrest. Epileptic seizures and obstructive sleep apnea (when breathing stops and restarts often during sleep) are causes of significant patient morbidity.

Data collection of time series:

Non-invasive and invasive methods. We need measurements to study the dynamics of physiological changes, which should provide a tabulation of the variables as a function of time. The time scales of measurements range from seconds to days. (there are also longitudinal studies).

An essential prerequisite for enhanced treatment of brain disorders is physiological data that can facilitate the proposition of testable hypotheses and the design of testable models of the brain and the mind. Thus, it is no wonder that millions of rodents, mammals, and primates are sacrificed in the quest for data on brain physiology and its complex electro-chemical structure. However, with increased awareness of both animal and human rights and formulated regulations and standards, using animals to collect experimental data is increasingly challenging. More importantly, there have been occurrences in which animal models are used for drug discovery, but the developed drugs prove ineffective in humans [37].

Time series analysis:

The vast amount of neural data obtained by sophisticated methods has opened a new era of brain research. New data analysis methods are being developed to exploit the available data fully. Seizure zones, for example, were traditionally localized manually, using the human brain's perfect pattern matching or mismatch recognition skills to identify the first pathological patterns at initiating epileptic seizures. Today, mathematical methods can help automate this detection and examine possible markers of epileptic tissue, either ictally (i.e., during seizures) or interictally (i.e., between seizures).

A now often-studied problem is uncovering the causal relationship among brain regions. Interestingly, most causality analysis methods are based on Norbert Wiener's principle on predictability: a time series is said to be causal to another if its inclusion makes the prediction of the caused time series more precise. (Remember Wiener's prediction: 1.1.1.1). The first practical and applicable implementation of this principle is Granger causality, introduced by Clive Granger (1934–2009) in 1969, who got the Nobel Prize in economics for developing and applying his method. For a short review, see [38].

3.4.2 Translating Mathematics into Medical Practice

Translating mathematical approaches into medical practice is a big challenge. A workshop on *Dynamical Disease–from the Blackboard to the Bedside* was organized in Montreal in November 2019, and the meeting demonstrated the progress that happened in the last 25 years [39]. Mathematical models for specific physiological (from hormonal, via cardiac to neural) systems were studied extensively. As conventional modeling techniques were supplemented by data collecting and processing devices and algorithms, combining the "model-driven" and "data-driven" perspectives has been time.

3.4.2.1 Sudden Deaths: A Failure of Feedback Control

Sudden infant death syndrome (SIDS), sudden cardiac death (SCD), and sudden unexpected death in epilepsy (SUDEP) are the most frequent classes of what is called sudden death. Recent studies state that SIDS occurs due to a failure of feedback control [40].

Respiratory failure is the primary cause of death. Acid reflux is potentially dangerous during a seizure, and several risk-reducing strategies were identified. Oxygen-conserving reflexes may have a role in sudden death. In this case, they are likely hyperactivated due to feedback and control systems failing to respond to unusual situations appropriately. If these reflexes are hyperactivated, death may be caused by a feedback loop maintaining a robust central apnea. (Sleep apnea is a common condition in which one's breathing stops and restarts many times while one sleeps. This can prevent her body from getting enough oxygen.)

Neuroengineers have developed devices and techniques in the last decades. Studying the underlying anatomical structures, neurophysiological mechanisms, and engineering possibilities is necessary. Many neuroengineering devices indeed benefit patients even before their mechanism is fully understood. Pacemakers, deep brain stimulators, and vagal nerve stimulators are a few known examples. However, understanding the operation of control loops is indispensable to making meaningful progress in preventing SIDS.

3.4.2.2 Predicting and Controlling Epilepsy

One of the primary goals of healthcare is the early identification and prediction of a disease to give timely preventative measures. This is especially true for epilepsy, a condition marked by repeated and somewhat unpredictable seizures. Patients can be spared the negative repercussions of epileptic seizures if they can be predicted in advance. Seizure prediction remains an unsolved subject despite decades of research. This will likely continue partly due to a lack of data to solve the problem. New advancements in machine learning-based algorithms can potentially bring a paradigm change in the early and accurate prediction of epileptic seizures [41].

Closed-loop control of epilepsy is suggested to be possible by transcranial electrical stimulation [42]. It is now well-documented that many neurological and psychiatric diseases are associated with clinically detectable, altered brain dynamics. At least in principle, dynamic brain activity can be restored by appropriate electrical stimulation. In epilepsy, abnormal patterns occur intermittently. Therefore, closed-loop feedback brain control that leaves other aspects of brain function intact is desirable. Here, we show that seizure-induced transcranial feedback electrical stimulation (TES) can significantly reduce spike episodes in a rodent model of generalized epilepsy. Closed-loop TES can be a powerful clinical tool for alleviating pathological brain patterns in drug-resistant patients. The real-time control of epilepsy is a hot research topic.

3.4.2.3 Therapeutic Strategies

Evidence-based medicine: Medical professionals make decisions based on the best available scientific concepts and data. *Evidence-based medicine* is now the generally accepted best practice. The romantic character of the humanistic, experienced old doctor with a white beard is still with us, but data and algorithms are good friends for the patients. Evidence-based medicine attempts to provide clinical benefits of tests and treatments using various mathematical methods, from statistical analyses to dynamic modeling.

Risk assessment and management: Cancer is a class of diseases characterized by out-of-control cell growth. Cancer growth and invasion happen across multiple time- and spatial scales; therefore, *multi-scale modeling* technique has begun to play a more critical role in transferring mathematical methods toward clinical implementations. Cell signaling mechanisms, the natural regulators of biological systems, are usually investigated at *molecular* scale. *Microscopic* models describe the malignant transformation of normal cells and associated alterations of cell-cell and cell-matrix interactions. Macroscopic-scale models deal with length scales of millimeters to centimeters and timescales of days to years. They grasp the gross tumor behavior dynamics, including morphology, shape, and extent of invasion [43]. Specifically, stem cell dynamics can be used to validate the number of blood-producing stem cells as a biomarker for risk assessment in patients who suffer from acute myeloid leukemia. The basic idea is that leukemic stem cells out-compete healthy cells. Thus, a lower healthy cell number implies more severe disease [44].

From algorithm to neurosurgery: For those epileptic individuals whose seizures can not be controlled by anticonvulsant medications, surgical removal of the epileptic focus can be life-changing. Models of cortical networks are the basis of stimulating the effect of surgical interventions and finding optimal surgery operations to get the best possible seizure reduction [45].

3.5 Ecological Systems

Ecological systems contain many negative feedback loops that keep parts of the systems within the boundaries necessary for the whole system to function. Negative feedback loops between predators and prey ensure that animal and plant populations are kept within the carrying capacity limits. (The carrying capacity of an environment is the *maximum population size* of a biological species that that specific environment can sustain.)

Positive feedback destabilizes the status quo, and it is a driving force for changes and steers the system outside of its normal operation range. As we saw earlier, more population leads to more births and an increasing population. Sometimes, negative and positive feedback compensate for each other, but occasionally one becomes dominant. As we also already know, bistability is a frequent feature of complex systems. The succession of an ecosystem from grass to a shrub community is a well-studied example. While negative feedback keeps the biological community the same, a positive feedback process supports the idea that shrubs take over the environment relatively quickly.

As we remember in 1.1.2, vicious circles are a chain of events implying negative results. The melting of polar ice caps is an example. Ice is white and very reflective, unlike the dark ocean surface, which absorbs heat faster. As the atmosphere warms and sea ice melts, the darker ocean absorbs more heat, causes more ice to melt, and makes the Earth warmer overall. The ice-albedo feedback is a powerful illustration of positive feedback. Virtuous circles, on the other hand, are a chain of events that reinforce themselves through a positive feedback loop, creating favorable outcomes.

Human ecology examines the results of human exercises as a chain of impacts through the environment and human social framework. The Reader may remember the DDT Story 1.2.3.2. It had a malignant effect on animals and humans, but...we cannot forget that it also saved millions of lives.

A well-studied case is related to a popular activity: fishing. Fishing is coordinated toward one portion of the marine environment, specifically the fish, but fishing has unintended impacts on other parts of the biological system. Those impacts set in movement an arrangement of the extra effects that go back and forth between the environment and social framework [46].

Drift nets are nylon nets that are undetectable within the water. Fish ended up tangled in float nets while attempting to swim through them. In the 1980s, anglers utilized thousands of kilometers of drift nets to capture fish in oceans worldwide. Within the mid-1980s, it was found that drift nets were murdering vast numbers of dolphins, seals, turtles, and other marine creatures that suffocated after getting entrapped within the nets–an exchange of data from the biological system to the social framework, as portrayed in Fig. 3.7.

When conservation organizations realized what the nets were doing to marine animals, they campaigned against drift nets, mobilizing public opinion to pressure governments to make their fishermen stop using them. The governments of some nations did not respond, but others took the problem to the United Nations, which passed a resolution that all nations should stop using drift nets. At first, many fishermen did not want to stop using drift nets, but their governments forced them to change. The fishermen switched from drift nets to long lines and other fishing methods within a few years. Long lines, which feature baited hooks hanging from a mainline often kilometers in length, have been a standard method of fishing for many years. The long lines fishermen use to put several hundred million hooks in the oceans worldwide.

The drift net story shows how human activities can generate a chain of effects that passes back and forth between social systems and ecosystems. Fishing affected the ecosystem (by killing dolphins and seals), which led to a change in the social system (fishing technology). It is a never-ending story.

Environmental and social trends suggest that natural disasters, such as ecological extinctions or social collapses, may occur due to the impairment of the control system implemented by feedback loops.

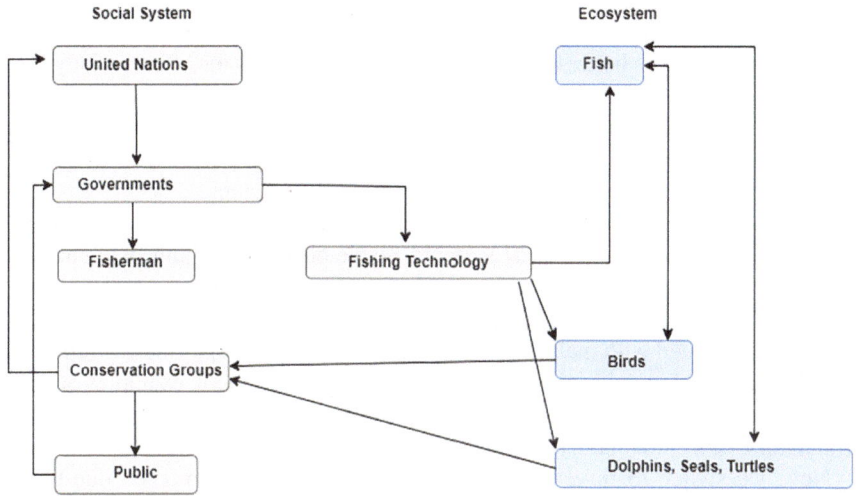

Fig. 3.7 Causal loops in the commercial fishing system [46]

3.6 Lessons Learned and Looking Forward

One of the most critical concepts in theoretical biology is homeostasis. Living structures are not in equilibrium with their environment, and their internal state is maintained by material, energy, and information flow. Feedback control is a fundamental tool at every level of the biological hierarchy, from cellular to socio-ecological systems.

Cellular function is also influenced by its environment. Adaptation to specific environments is performed by regulating the gene expression that encodes the enzymes and proteins needed for survival in a particular environment. Factors influencing gene expression include nutrients, temperature, light, toxins, metals, chemicals, and signals from other cells. Malfunctions in the regulation of gene expression can imply various disorders and diseases.

Biological clocks, organisms' natural timing devices, regulate the cycle of circadian rhythms. It provides a rhythmic function to our sleep-wake cycle. There is a master clock in the brain, which coordinates all the biological clocks in a living thing, keeping the clocks in synchronization.

Thermoregulation, a homeostatic mechanism, maintains physiologic body temperature by compensating heat generation with heat loss. Organisms can keep their body temperature within certain boundaries, even when the surrounding temperature is very different. A textbook example of negative feedback is the glucose regulatory system. When blood glucose concentration increases, the pancreas beta-cells secrete more insulin. Our ancestors had to make instantaneous decisions under terrifying threats, whether it was better to fight or flee. Stress represents the effects of anything that seriously threatens homeostasis, and control mechanisms help to provide survival. We discussed dynamic diseases, i.e., abnormal behavior due to the impairment of the control system. Neurological disorders are the best-studied cases. Ecological crises also may happen when the control system is not efficient.

The message for the Reader is that feedback control strategies maintain the stable, healthy operation of biological systems. Stable may refer to both stationary and oscillatory, clock-like systems. Self-organization and self-regulation are the basis of biological functioning. Internal control mechanisms ensure this functioning. Many disorders associated with malfunctioning emerge due to the impairment of this control system. Pathological cardiac and neural rhythms are textbook examples of these disorders in temporal patterns. To restore order, external control (pacemakers, brain stimulators, neuromodulatory drugs, etc.) should be adopted.

Now, we switch to investigating natural disasters, which emerged due to the impairment in the control loop.

References

1. Huffman C (2021) "Alcmaeon", the stanford encyclopedia of philosophy (Summer 2021 Edition)
2. Billman GE (2020) Homeostasis: The underappreciated and far too often ignored central organizing principle of physiology. Front Physiol 10. https://www.frontiersin.org/articles/10.3389/fphys.2020.00200/full
3. Ribatti D (2009) William Harvey and the discovery of the circulation of the blood. J Angiogenes Res 2009(1):3
4. Schultz SG (2002) William Harvey and the circulation of the blood: The birth of a scientific revolution and modern physiology. Physiology 17(175–180) https://journals.physiology.org/doi/full/10.1152/nips.01391.2002
5. Cooper SJ (2008) From Claude Bernard to Walter Cannon. Emergence of the concept of homeostasis. Appetite 51(419–427) https://www.sciencedirect.com/science/article/pii/S0195666308005114
6. Holmes FL (1986) Claude Bernard. The Milieu Intérieur, and regulatory physiology. Hist Phil Life Sci 8(3–25)
7. Cannon WB (1963) The wisom of the body. W. W. Norton & Company; Rev Enl Ed edition
8. Walter BC https://www.whonamedit.com/doctor.cfm/3178.html
9. Érdi P, Szvetelszky ZS (2022) Repair: When and how to improve broken objects, ourselves, and our society. Springer
10. Lovelock JE, Margulis L (1974) Atmospheric homeostasis by and for the biosphere: The gaia hypothesis. Tellus 26(1–2):2–10
11. Wilson D, Marcschinsky FM (2021) Metabolic homeostasis in life as we know it: Its origin and thermodynamic basis. Front Physiol 12:658997
12. Watson AJ, Lovelock JE (1983) Biological homeostasis of the global environment: The parable of Daisyworld. Tellus B 35(4):286–9
13. Tyrell T, Gaia: the verdict is ... (2013) The new scientist 26 October 2013, pp 30–31
14. Tyrell T (2013) On Gaia: A critical investigation of the relationship between life and earth. Princeton University Press
15. Inducer (2023) Wikipedia. Retrieved, May 11
16. Jacob F, Monod J (1961) Genetic regulatory mechanisms in the synthesis of proteins. J Mol Biol 3:318–356
17. Monod J, Jacob F (1961) Telenomic mechanisms in cellular metabolism, growth and differentiation. Cold Springer Harbor Symp. Quant Biol 26(389–401)
18. Morange M (2005) What history tell us. J Biosci 30(313–316)
19. Yildirim N, Mackey MC (2003) Feedback regulation in the lactose operon: A mathematical modeling study and comparison with experimental data. Biophys J 84:2841–2851
20. Novick A, Weiner M (1957) Enzyme induction as an all-or-none phenomenon. Proc Natl Acad Sci USA 43:553–566
21. Robeva R, Yildirim N (2013) Bistability in the lactose operon of Escherichia coli: A comparison of differential equation and Boolean network models. In: Mathematical concepts and methods in modern biology, using modern discrete models, pp 37–74
22. Santillán M, Mackey MC, Zeron ES (2007) Origin of bistability in the lac operon. Biophys J 92(11):3830–3842. https://doi.org/10.1529/biophysj.106.101717
23. https://www.nobelprize.org/prizes/medicine/2017/press-release/
24. Érdi P (2007) Complexity explained. Springer
25. Chakravarty S, Hong CI, Csikász-Nagy A (2023) Systematic analysis of negative and positive feedback loops for robustness and temperature compensation in circadian rhythms. NPJ Syst Biol Appl 9:5. https://www.ncbi.nlm.nih.gov/pmc/articles/PMC9922291/
26. Novak B, Tyson JJ (2008) Design principles of biochemical oscillators. Nat Rev Mol Cell Biol 9(12):981–991
27. Kuramoto Y (1984) Chemical oscillations, waves, and turbulence. Springer-Verlag, New York, NY

28. Sarfati R, Hayes JC, Peleg O (2021) Self-organization in natural swarms of *Photinus carolinus* synchronous fireflies. Sci Adv 7(28)
29. Hartbauer M, Haitzinger L, Kainz M, Romer H (2014) Competition and cooperation in a synchronous bushcricket chorus. Roral Soc Open Sci 01 October 2014
30. Strogatz S (2000) From Kuramoto to Crawford: Exploring the onset of synchronization in populations of coupled oscillators. Physica D 143(1–4):1–20
31. Knechtle B, Waśkiewicz Z, Sousa CV, Hill L, Nikolaidis PT (2020) Cold water swimming-benefits and risks: A narrative review. Int J Environ Res Public Health 17(23):2
32. How the fight or flight response works https://www.stress.org/how-the-fight-or-flight-response-works
33. Understanding the stress response https://www.health.harvard.edu/staying-healthy/understanding-the-stress-response
34. Fight F, Freeze F, Responses to stressful situations https://www.masterclass.com/articles/fight-flight-freeze-fawn#5mggTKqApFjXT5ESuoFZth
35. Hofmann SG, Asnaani A, Vonk IJ, Sawyer AT, Fang A (2012) The efficacy of cognitive behavioral therapy: A review of meta-analyses. Cognit Ther Res 36(5):427–440
36. Belair J, Glass L, An Der Heiden U, Milton J (1995) Dynamical disease: Identification, temporal aspects and treatment strategies of human illness. Chaos 5(1):1–7
37. van der Worp HB, Howells DW, Sena ES, Porritt MJ, Rewell S, O'Collins V, Macleod MR (2010) Can animal models of disease reliably inform human studies? PLoS Med 7(3):e1000245
38. Zsigmond B, Dániel F, Zoltán S (2017) In time series and interactions: Data processing in epilepsy research. In: Péter É, Basabdatta SB, Amy LC (eds) Computational neurology and psychiatry. Springer, pp 73–91
39. Bélair J, Nekka F, Milton JG (2021) Introduction to focus issue: Dynamical disease: A translational approach. Chaos 31(6):060401
40. Budde R, Biggs E, Irazoqui P (2023) Sudden deaths: A failure of feedback control. In: Thakor NV (ed) Handbook of neuroengineering. Springer, Singapore
41. Rasheed K, Qayyum A, Qadir J, Sivathamboo S, Kwan P, Kuhlmann L, O'Brien T, Razi A (2021) Machine learning for predicting epileptic seizures using EEG signals: A review. IEEE Rev Biomed Eng 14:139–155
42. Berényi A, Belluscio M, Mao D, Buzsáki G (2012) Closed-loop control of epilepsy by transcranial electrical stimulation. Science 337(6095):735–7
43. Deisboeck TS, Wang Z, Macklin P, Cristini V (2011) Multiscale cancer modeling. Annu Rev Biomed Eng 15(13):127–55
44. Stiehl T (2020) Using mathematical models to improve risk-scoring in acute myeloid leukemia. Chaos 30:123150
45. Junges L, Woldman W, Benjamin OJ, Terry JR (2020) Epilepsy surgery: Evaluating robustness using dynamic network models. Chaos 20:113106
46. Marten GE (2001) Human ecology–basic concepts for sustainable development. Earthscan Publications

Chapter 4
Climate Changes, Wildfires, Tsunamis

Abstract This chapter starts with analyzing the role of positive and negative feedback loops in climate systems. Self-reinforcing positive feedback loops could result in an irreversible tipping point when climate spins out of control, with catastrophic results. Reducing the chance of climate catastrophe must become a central focus for civilization today, and appropriate feedback control strategies should be implemented.

Natural disasters may occur due to impairments within a feedback control system. As was mentioned in Sect. 3.5, an example of a positive feedback loop in the climate system is the ice-albedo effect: less sea ice due to warming results in a darker sea surface overall, which reflects less heat, raises local temperatures, and leads to yet more sea ice loss. A tipping point is hit if negative feedback does not stop this process, and a significant shift to a "new normal" state becomes inevitable. In this chapter, we will discuss more details since understanding the mechanisms leading to global warming and its possible control strategies is crucial for the future of our civilization.

There has also been an increase in the frequency and severity of heat waves, droughts, wildfires, storms, tsunamis, and floods. Accumulated data, model studies, and arguments convincingly suggest that every aspect of life should be considered in terms of decreasing carbon footprints.

4.1 Feedback Loops and Climate Changes

4.1.1 The Climate System

The climate is an excellent example of a complex system. It contains five subsystems: atmosphere, hydrosphere, cryosphere, biosphere, and the Earth's crust. The interaction among these subsystems is one source of climate dynamics.

Climate change and energy budget: The weather gets energy from the Sun and a little from the Earth's center and the Moon's tides. The Earth sends energy to outer space in two ways: it reflects some of the Sun's energy and also gives off its own energy as infrared radiation. The amount of energy going in and out and how it

© The Author(s), under exclusive license to Springer Nature Switzerland AG 2024   55
P. Érdi, *Feedback*,
https://doi.org/10.1007/978-3-031-62439-1_4

moves through the climate system determines how much energy the Earth has. When the amount of energy coming into Earth is more than the energy going out, Earth's energy budget becomes positive, and the climate system starts to warm up. If more energy is left, the Earth gets colder.

Changes caused by the system's five components and dynamics are called *internal* climate variability. The system is also influenced by *external* forcing phenomena from outside the system. The following are the most critical external forcing phenomena:

- Incoming sunlight
- Greenhouse gases
- Aerosols
- Land use and cover change

Incoming sunlight: Sun energy varies over time; one reason is the circa 11-year solar cycle. Greenhouse gases, such as carbon dioxide, methane, and nitrous oxide, trap heat in Earth's atmosphere. While they let sunlight pass through the atmosphere, they prevent the heat that the sunlight brings from leaving the atmosphere. Carbon dioxide is the most dangerous greenhouse gas. We (almost) all know that human activity, specifically burning fossil fuels like coal and oil, is the primary source of the dramatic increase in the concentration of CO_2. Additionally, the effect of methane cannot be underestimated; it is 25 times more effective in trapping heat than CO_2, and its quantity in the atmosphere has more than doubled since the beginning of the Industrial Revolution.

Aerosols: The atmosphere contains liquid and solid particles. Natural aerosol sources exist from the hydrosphere, the Earth's crust, and meteorites. Particles from fossil fuel combustion or wildfire are products of human activity.

Changes in the cover of the land might have effects on the climate. Recent studies found "unsustainable human use dried up lakes such as the Aral Sea in Central Asia and the Dead Sea in the Middle East, while lakes in Afghanistan, Egypt, and Mongolia were hit by rising temperatures, which can increase water loss to the atmosphere" [1]. Rising sea levels and deforestation also imply changes in the cover, so the changed cover can capture more or less sunlight. (There is an online platform, Global Forest Watch (GFW) [2], that provides data and tools for monitoring changes in forests. You can receive actual information about changes in the forests worldwide. The total area of humid primary forests decreased globally by seven percent in the last twenty years.)

Feedback loops can either amplify or diminish the effects of climate *forcings*. The different components of the climate system respond to external forcing with very different velocities. The response time of the atmosphere can be expressed on the timescale of days or so, while the deep layers of the hydrosphere, or the cryosphere, are much slower and can be measured at the scale of centuries.

The Intergovernmental Panel on Climate Change (IPCC) report on the physical climate processes and feedback describes the state of the art of the scientific perspective [3, 4].

4.1.2   Positive Climate Feedback

Water vapor feedback: If the air gets warmer, it can hold more water vapor, so there will be more water vapor in the air. Water vapor is a type of gas in the atmosphere that can cause a warming effect. When there is more water vapor in the atmosphere, it makes the air even warmer. This warming then leads to more water vapor in the air (positive feedback). This cycle continues until other factors stop it from happening. The outcome is a bigger greenhouse effect than what is caused by only CO_2.

Cloud feedback: Recent advances in collecting data using satellites provide new information about the role of clouds in emitting and trapping heat. Clouds can form in different parts of the sky when humid air rises and cools down, causing the water vapor to turn into water droplets or ice crystals. Clouds are made of tiny bits of water or ice, which can come together to make rain, snow, or drizzle. These tiny processes work together with particles in the air, sunlight, and how the air moves around to create clouds. This highly complex process happens in many different sizes and periods. We cannot go into the details here, but the estimation of the cloud feedback (and, of course, of all the other subprocesses) is very complicated. Clouds should be decomposed into regimes, such as the altitude level at which each type is usually found. This is the language that ICPP uses: "Multiple lines of evidence (theory, observations, emergent constraints, and process modeling) are now available in addition to Earth Systems Models simulations, and the positive low-cloud feedback is consequently assessed with high confidence [4]."

Ice–albedo feedback: Near the poles, the temperature increase is more significant than in the planetary average. Polar warming is more dramatic than tropic warming. There is a simple self-amplifying mechanism: water is less reflective than white ice (corresponding to a high albedo) and thus absorbs more solar radiation. This causes more warming, which in the next step causes more melting, and the cycle continues.

Carbon-release feedback: When the Earth gets hotter, the frozen soil melts. This frozen ground holds a lot of carbon trapped in the soil. So when the permafrost (defined as any ground that remains completely frozen for at least two years) thaws, it releases methane gas into the air. When these gases are released, they make the Earth warmer. This warmth makes more ice melt, which consequently releases more carbon.

4.1.3   Negative Climate Feedback

Blackbody radiation (also called Planck feedback): When an object gets hotter, it gives off more energy. Planck feedback is a powerful force that has a stabilizing effect. Think about an object in a state where the amount of radiation energy it absorbs is the same as the amount of radiation energy it releases. Now, imagine that the amount of radiation energy coming in becomes less. The object is not balanced anymore because it gets less energy than it gives. When this happens, the temperature of the object will go down. This reduces the radiation energy released, and the system returns to a balanced radiation state. This is a situation in which the system prevents temperature change instead of causing it to increase. This also applies when more energy is put in. If the system is already in balance with the amount of energy it absorbs and releases, we increase the amount of energy it absorbs. The system's temperature will increase because more energy is coming in than going out. The temperature keeps going up until it reaches a new state of balance. Once again, this is a negative feedback response because the sequence of events restricts the temperature change instead of making it more robust.

Chemical weathering: There are other processes which remove CO_2 from the atmosphere. Chemical weathering acts on a long time scale. Global warming increases weathering and implements feedback from the climate to Earth's surface.

Synthesis of organic compounds from CO_2: When plants perform photosynthesis, CO_2 is combined with water and solar energy and converted to carbohydrates. A more significant CO_2 concentration increases photosynthesis's velocity, i.e., of carbon dioxide removal.

A famous figure from Al Gore's book *An Inconvenient Truth:* [6] (Fig. 4.1) contains most of the significant positive and negative feedback loops known in the early years of the 21st century.

4.2   The Truth Is Still Inconvenient for Many Stakeholders

4.2.1   Climate Change: Myth or Reality?

The starting point of this book is that a narrow border separates prosperity from catastrophes, and a fine-tuned control problem is to keep human civilization from destruction. There are recurring debates about whether global warming (i) is real or a myth, (ii) if it is real, is it a product of human activity, and (iii) if it has more negative or positive effects.

   Global warming is real. Global warming (or, using a more neutral and more general expression, climate change) has been the subject of much political controversy, mainly in the United States and other countries. Climate change is identified with rising temperatures and other extreme events, such as higher sea levels, via tsunamis and wildfires. Scientific evidence supports the view that the quantity of heat-trapping greenhouse gases in the atmosphere is increasing.

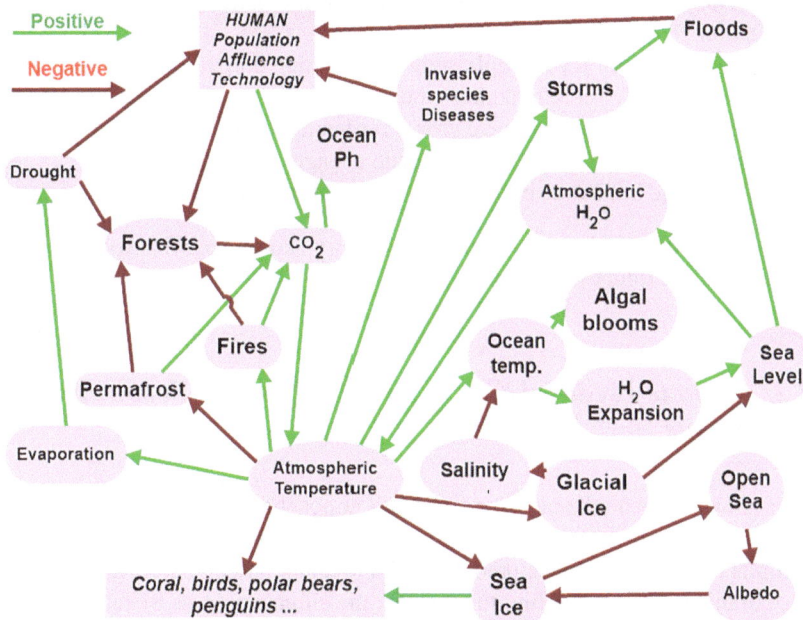

Fig. 4.1 Global climate change is mostly caused by humans releasing more carbon dioxide into the air. This leads to a complex system that can worsen or improve the effects. Some examples of these effects include extreme heat, floods, droughts, and new diseases. All of these things end up having a bad impact on the number of people in the world

As an analysis of the National Geographic [7] states: "The answer to the question *Is global warming real?* is yes: Nothing other than the rapid rise of greenhouse gas emissions **from human activity** can fully explain the dramatic and relatively recent rise in global average temperatures." There is another question: are there positive benefits from global warming? The research suggests that crops and plants generally do better when more carbon dioxide is in the air, which also helps them withstand dry conditions. However, this advantage has a downside: unwanted plants, like invasive species and bugs that harm crops, will also do well in a hotter environment. Water will be less available in areas without much rain and needing moisture for farming. Eventually, the side effects of having more carbon dioxide for crops will probably be overtaken by the negative consequences of heat stress and insufficient water [8].

When considering the possibility of controlling climate change, knowing how people feel about it is helpful. For example, a recent analysis of the Pew Research Center (a nonpartisan fact tank) about What the data say about **Americans' views of climate change** [10]:

- A majority of Americans support the U.S. becoming carbon neutral by 2050
- Americans are reluctant to phase out fossil fuels altogether, but younger adults are more open to it

- The public supports the federal government incentivizing wind and solar energy production
- Americans see room for multiple actors—including corporations and the federal government—to do more to address the impacts of climate change.
- Democrats and Republicans have grown further apart over the last decade in their assessments of the threat posed by climate change
- Climate change is a lower priority for Americans than other national issues.

However people feel about climate change, we should see what today's science tells us about the future of climate.

4.2.2 Climate Predictability: A Short Note About Scope and Limits

As it is known from nonlinear science, Earth's atmosphere-ocean dynamics behavior is *chaotic*, i.e., it shows sensitivity to slight perturbations in initial conditions. This fact limits our ability to make predictions for weather and climate. Another natural source of this limit is our ability to represent the climate process by faithfully using physics-based mathematical models.

To improve the prediction quality due to uncertainties both in the model structure and initial conditions, simulations are repeated many times from different initial conditions and using different, somewhat competitive models. Climate ensemble simulations are used to study the *theory of parallel climate realizations*, which states that "it is worth imagining many replicas of the Earth System that evolve in parallel, but differently, although they all are subjected to the same physical laws and time-dependent set of forcing and boundary conditions" (e.g., in terms of irradiation and greenhouse gas concentration) [11].

News from "today" (July 6 2023 [5])
Heat Records Are Broken Around the Globe as Earth Warms, Fast
From north to south, temperatures are surging as greenhouse gases trap heat in the atmosphere and combine with the effects of El Niño. The past three days were likely the hottest in Earth's modern history, scientists said on Thursday, as an astonishing surge of heat across the globe continued to shatter temperature records from North America to Antarctica.

The spike comes as forecasters warn that the Earth could be entering a multi-year period of exceptional warmth driven by two main factors: continued emissions of heat-trapping gases, mainly caused by humans burning oil, gas, and coal, and the return of El Niño, a cyclical weather pattern.

4.2.3 *From Al Gore to Greta*

The former U.S. vice president, Al Gore, is a decorated early activist against unwanted climate change generated by human activity. His book and documentary (won the Academy Award for Best Documentary Film) entitled *An Inconvenient Truth* led him to receive the Nobel Peace Prize in 2007, along with the IPCC, for his activity in promoting action against climate change. His mother generated his interest in environmental issues in his teenage years by showing him Rachel Carson's Silent Spring (as the Reader remembers, this super-influential book was mentioned in the Sect. 1.2.3.2).

It would be difficult to deny that the movie raised public awareness of climate change. We all know that even the most benevolent actions may have unintended consequences. Since Al Gore is known as a very prominent Democrat politician, his views catalyzed the growing polarization of American public opinion on climate change [9].

Social media platforms provide essential locations for the everyday discussion and debate of climate change. While the initial hope was that social media communication would contribute to democratization, it looks like it is accelerating political polarization [12].

How Do We Address the Polarization Around Climate Change? [13]

According to statistics from 2022, 24% of Americans continue to believe that human activity plays no part in climate change. To make progress in enacting regulations and limiting emissions, we need much more cooperation to end our dependence on fossil fuels. How can we reduce polarization both at individual and institutional levels? First, we should realize that mutual repetition of the previous message louder and louder materializes an uncompensated positive feedback loop leading to an uncontrollable "explosion".

Feedback control: a conversation with the other side. Conflicts between groups cannot be resolved in the long term unless the motivational conditions from which conflicts arise are eliminated or reduced. Peter T. Coleman, a professor of psychology at Columbia University and executive director of the Columbia Climate School's Advanced Consortium on Cooperation, Conflict, and Complexity, studied conflict and polarization. He agrees with people who believe that two-thirds of Americans belong to the *Exhausted Majority*. The members of this significant social group are fed up with the polarization of society and hope to find common ground. It is time to support them in speaking more loudly to provide a stabilizing negative feedback mechanism as a controlling response to the unlimited mutual positive feedback of the opposite views.

I see the effects of two super activists due to a social feedback control. Both of them perceived the difference between the actual and the required state of the environment and made steps to reduce the deviation.

Al Gore is undoubtedly a leading voice. He discussed climate change as an issue as early as 1976 (he was 28 years of age) in the United States Congress. As a vice president, he was the top supporter in the U.S. to take action in the spirit of the Kyoto Protocol (1997), the first international attempt to restrain climate change. While it happened that he did not become the president of the USA, I do not believe we can underestimate his activism for environmental and climate problems. I see the pouting lips of some people stating that his results are paltry and occasionally he used private jets [14].

Greta has been the target of a flow of disinformation and conspiracies. I wrote Greta and not Great Thunberg. (When my parents mentioned Greta without her family name, in their mind, there was another Swedish lady, the melancholic movie star Greta Garbo (1905–1990)). In 2018, she initiated weekly protests against climate inaction outside the Swedish Parliament (Fridays for Future). The demonstration became a global movement involving millions of students across more than 150 countries. Through her protests and speeches, she has unified the world about the climate crisis in probably more efficient ways than Al Gore's.

Climate skeptics label her as a spoiled child, a leftist pawn just following instructions from George Soros. While social media is the primary source of these attacks, far-right activists, politicians, and even heads of state, including Trump, played a role.

Greta seems to have moral power. Her activism has its roots in science rather than politics. Unlike other high-profile climate activists, she can't easily be accused of occasional hypocrisy: Besides being vegan, she abstains from plane travel and mass consumerism. Her undeniable influence on the general public, as well as on politicians and maybe even on corporations, is called the "the Greta effect". However, her participation in the Gaza movement could pose a problem [15] for the climate change movement.

A few big influencers might have a tremendous effect on controlling public opinion. It will be fascinating to see the future debate when nobody can be a neutral passive observer. There is also a spectrum of climate skeptics, from the uneducated to the most educated. The most respected scientist who turned out to be a climate skeptic was Freeman Dyson (1923–2020). For a balanced analysis of his views, see [16]:

> When Dyson joins the public conversation about climate change by expressing concern about the "enormous gaps in our knowledge, the sparseness of our observations, and the superficiality of our theories," these reservations come from a place of experience. Whatever else he is, Dyson is a good scientist; he asks the hard questions. He could also be a lonely prophet. Or, as he acknowledges, he could be dead wrong.

4.3 Wildfires

4.3.1 The (Not Only) Canadian Wildfire in 2023

What Do the Data Tell? A wildfire is an unplanned, unpredictable, and uncontrolled burn in naturally ignitable vegetation. Forests, grasslands, and prairies are the most common wildfire zones. While this summer (2023), the Canadian wildfires got considerable publicity, the story is not new. The list of the top 12 largest historical wildfires and the damage they caused to biodiversity, ecosystems, and urban settlements is as follows [17]:

1. 2003 Siberian Taiga Fires (Russia)—55 Million Acres
2. 2019/2020 Australian Bushfires (Australia)—42 Million Acres
3. 2014 Northwest Territories Fires (Canada)—8.5 Million Acres
4. 2004 Alaska Fire Season (US)—6.6 Million Acre
5. The Great Fire Of 1919 (Canada)—5 Million Acres
6. 1950 Chinchaga Fire (Canada)—4.2 Million Acres
7. 2010 Bolivia Forest Fires (South America)—3.7 Million Acres
8. 1910 Great Fire of Connecticut (US)—3 Million Acres
9. 1987 Black Dragon Fire (China and Russia)—2.5 Million Acres
10. 2011 Richardson Backcountry Fire (Canada)—1.7 Million Acres
11. The 1989 Manitoba Wildfires (Canada)—1.3 Million Acres

In the last few years, there has been a growing trend in the number of wildfires, causing increasing problems. One of the most significant events of our time is the Canadian wildfires. Canada has been wrestling with an extraordinary wildfire season in 2023. Starting in March and intensifying in June, the country has witnessed a series of record-setting wildfires. Eleven provinces and territories have been affected, including Alberta, Nova Scotia, Ontario, and Quebec. As of July 9, 2023, 3,831 fires have burned 9,333,743 hectares. The scale of these wildfires has surpassed any previously recorded in Canadian and North American history. The Canadian Interagency Forest Fire Centre declared it the worst wildfire season on record.

Across the United States, nearly 7.5 million acres are lost to wildfires yearly, with a risk of damage to every state. Over a year, wildfires accounted for over $3.2 billion in damage. Areas in the west, like California, Nevada, and Arizona, characterized by arid climates and little rain, are especially susceptible to wildfires. They can be devastating, destroying not only individual homes but entire communities.

European Forest Fire Information System (EFFIS) has been established in 2000 [18]. As the annual report writes, "2021 was again a drastic year that raised our concern about wildfires and their impacts in the European Union and its neighbor countries. Over 5500 km^2 (1.359 million acres) of land were burned in 2021—more than twice the size of Luxembourg—with over 1000 km^2 (over 247 thousand acres) within protected areas of Europe's Natura 2000 network. The EU's reservoirs of biodiversity. 2022 was the second-worst wildfire season in the European Union. The damage caused in many of these invaluable ecosystems will take many years to restore." Most Mediterranean countries (including the North African ones) were affected.

Wildfires are a growing problem because they destroy property and impact air quality, agriculture, resources, transportation, and animal and human health.

Causes: Three ingredients, fuel, oxygen, and a heat source, are necessary to generate fire. Firefighters call these three elements the fire triangle. Brush, grasses, and trees are fuels. Air is the oxygen supply; burning campfires or cigarettes can be heat sources.

Wildfires might have many reasons [19]. First, climate change has contributed to warmer and drier conditions at the macro scale: vegetation has become more suscepti-ble to ignition. There is a *positive feedback* mechanism: "The wet gets wetter, the dry gets drier". Second, the development of communities changes the "Wildland-Urban Interface". With the reduced distance between urban areas and wild landscapes, there is a much higher risk that destructive wildfires will reach cities and towns. Third, forest management practices are somewhat poor, with the traditional emphasis on fire suppression. People who cut down trees and their friends in politics sometimes say that getting rid of more trees or burned logs will make fires less likely. They believe that if there is less wood for the fire to burn, it won't spread as quickly. However, this understanding is incorrect and does not align with strategic forest management. In regular logging, they usually chop down the most significant and strongest trees, which are best at withstanding fires. This means they are removing the forest's essen-tial parts that help prevent fires. It would be better to remove smaller trees instead, but this needs to be done cautiously with advice from scientists and the public. Even trees that are dead or appear dead still have a purpose. They help support different types of plants and animals and strengthen the forest against fires. It can be risky to get rid of them. In addition, at the microscopic level, lightning strikes and irrespon-sible human activities, such as discarded cigarette butts and abandoned campfires, are also factors. While wildfires are officially labeled natural disasters, only about 15 percent occur independently. The other cases result from human causes, includ-ing unattended camp and debris fires, discarded cigarettes, etc. Whether natural or man-made, three elements are necessary to generate wildfire: fuel, oxygen, and a heat source. Trees, grasses, brush, and even homes might be fuels. Air is the oxygen supply, and the Sun or some human activities may play the role of heat sources. It is clear now that the combination of drought and elevated heat combined with human irresponsibility leads to more frequent and intense wildfires, posing significant forest threats.

Health and Social Impacts: Poor air quality has led to the cancellation of many events in Montreal and many communities in Quebec, including *Fête nationale*—also known as St-Jean-Baptiste Day. The impact of these fires extends beyond Canada's borders. Smoke from the wildfires has caused air quality alerts and evacuations in Canada and the United States. Furthermore, the smoke has even crossed the Atlantic and reached Europe, reflecting the global consequences of such events.

More generally, wildfires have devastating impacts on the environment, causing damage to plant populations and displacing or killing numerous animals. Emer-gencies caused by evacuations, physical injury, mental health consequences, and, at worst, loss of human life are extensively covered in the media. Blocked roads and railway lines, cuts of electricity, mobile and land telephone lines, destruction of homes and industries, and the way of life of many communities. The different impacts reinforce each other: one of the survivors' most significant challenges after the wildfire was accessing safe and secure shelter. Survivors had to move among various rentals, hotels, or shelters. It is not surprising that cumulative stress during this period of housing instability created or dramatically increased physical health issues [20].

National Geography published an analysis *Here's how wildfires get started—and how to stop them* [21]. The results of fierce blazes amplify past fabric harm, as they deliver smoke that can hurt human well-being. Delicate particulate matter in rapidly spreading fire smoke poses a coordinated hazard to respiratory well-being, particularly for those with pre-existing conditions. Fierce blazes can cause discussed contamination levels to surpass those in intensely contaminated cities like Beijing. The contamination from rapidly spreading fires can travel long distances, affecting discussion quality in completely different districts. The costs of these fires are not as they were money related but, moreover, biological, with carbon outflows from the fires compounding worldwide warming and contributing to a horrendous climate input circle. The impacts of the dry season and fierce blazes on biological systems should be considered comprehensively, considering the differences between life forms and their reactions to unsettling influences. Whereas the impacts on plants are well considered, the effect on creepy crawlies, which play a vital part in timberland biological systems, is less known. Specifically, ants and wild bees are imperative for supplement reusing, seed dispersal, and fertilization. Burned environments appear more conducive to more subterranean insect and bee species than unburned environments. A few beneficial characteristics of these insects continue in burned ranges within the long term. Understanding how these creepy crawlies recoup after fierce blazes gives bits of knowledge into the general biological system recuperation.

4.3.2 Damage Control

Prevention is the first step. Prevention teams head off fires in risky locations by clearing bushes and digging ditches. The Reader may want to see a short video about the difficult work of a Spanish fire prevention brigade: [22]. Unfortunately, prevention has its limits. The recovery of ecosystems affected by fires is slow. It can take years or even decades. Drought and extreme heat exacerbate the decay of forest masses, making them more vulnerable to pests and fires. The absence of rain and rising temperatures contribute to arid soil conditions, hindering reforestation efforts. To mention a European example, in Spain, a geographically vulnerable country, the effects of climate change and increased aridity are particularly pronounced. The combination of drought and heat leads to more frequent and intense wildfires, posing significant forest threats.

Firefighters fight bursts by eliminating some ingredients of the triangle. One conventional strategy is to drench existing fires with water and splash fire retardants. In some cases, firefighters work in groups, regularly called hotshots, to clear vegetation from the area around a fire to contain and, in the long run, starve it of fuel. The coming about tracts of arrival are called firebreaks.

Firefighters may also utilize controlled burning to halt a fierce blaze, making reverse discharges. This strategy includes battling fire with fire. These prescribed—and controlled—fires evacuate undergrowth, brush, and litter from timberland, denying something else a seething fierce blaze of fuel.

While private and public recovery organizations spend billions of dollars in the US providing relief to wildfire survivors, researchers from the San Francisco area studied California wildfires' health and social impacts and the deficiencies in current recovery resources [20]. Many clients remained without the resources to be physically, emotionally, and financially stable. Since the wildfire seasons are worsening year to year, organizations should rethink their possible strategies to improve their ability to support wildfire survivors' health and social needs in the future.

Researchers in the Pacific Northwest Research Station [23] identified a positive feedback loop that has negative outcomes for ecological values and a negative feedback mechanism to break the loop and reduce wildfire hazards: (Figs. 4.2, 4.3).

Wildfires are not only caused by climate change. They also add to it. There is a **vicious feedback loop** between wildfires and climate change, as illustrated by Fig. 4.4.

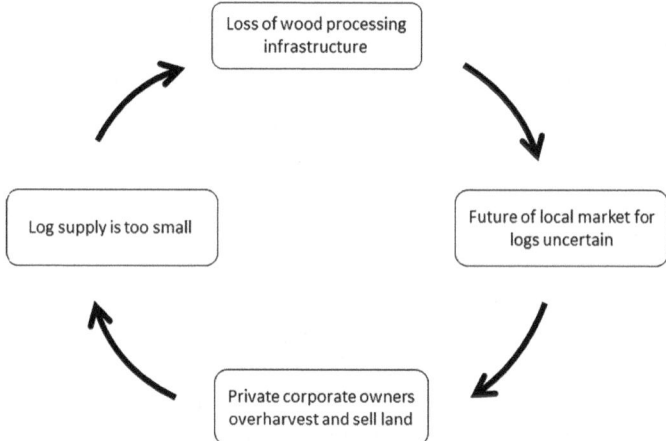

Fig. 4.2 Decline of local wood processing devices (mills) initiated a vicious loop

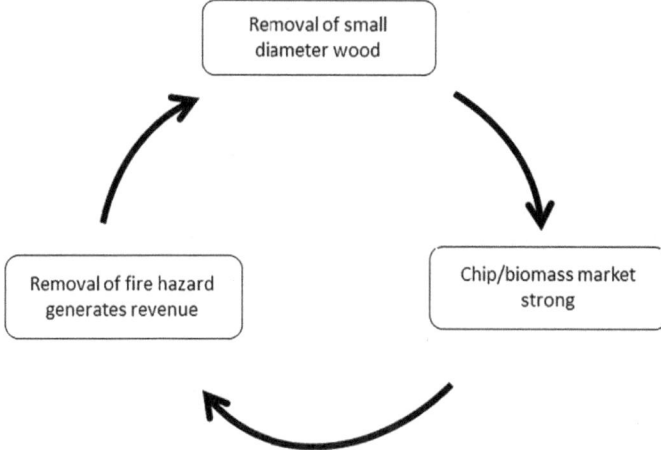

Fig. 4.3 The emergence of strong biomass markets can reduce wildfire hazard by making the removal of small-diameter trees and shrubs that need thinning economically possible

Fig. 4.4 Positive feedback between climate change and forest fires

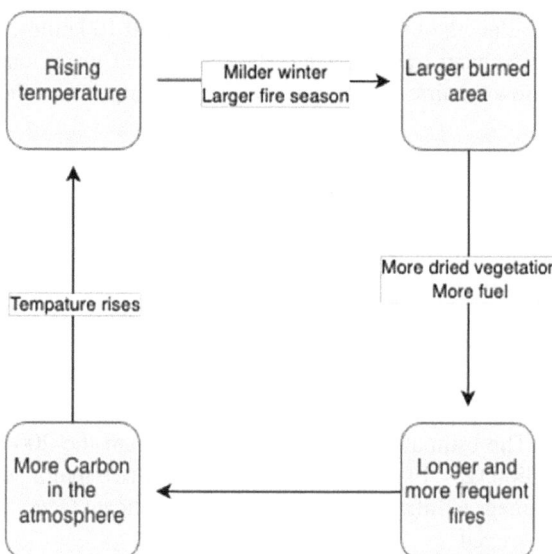

4.4 Tsunamis

The phenomenon

We learned about tsunamis, also called killer waves, on 26 December 2004. It is a Japanese word meaning harbor wave. A major earthquake with a magnitude of 9.1–9.3 M generated a massive tsunami with waves up to 30 m high on the surrounding coasts of the Indian Ocean. It was one of the deadliest natural disasters in recorded history, killing about 250,000 people in 14 countries.

The sudden vertical rise of the seabed by a few meters amid the seismic tremor uprooted gigantic volumes of water, coming about in a tidal wave that struck the coasts of the Indian Sea. A tsunami that causes harm distant from its source is sometimes called a teletsunami and is much more likely to be delivered by the vertical movement of the seabed than by flat action. The sudden vertical rise of the seabed by a few meters amid the seismic tremor uprooted enormous volumes of water, coming about in a torrent that struck the coasts of the Indian Sea.

How these huge waves are generated?

Earthquakes, landslides, and volcanic eruptions may generate tsunamis. Tsunami waves move rapidly across oceans. The speed and height of the tsunami wave depend on the depth of the ocean floor. At an ocean depth of around 20,000 feet, tsunami waves are less than a foot high, and their velocity is about 550 mph—about the speed of an intercontinental airplane. However, as the tide reaches shallower water, the speed of a tsunami wave is reduced while the height dramatically increases. The wavelength of the tsunami can reach about 100 miles. Large wavelength implies that the tsunami travels huge distances without losing only a little energy. Others may follow the first wave, and some may be larger than the previous ones [24].

Environmental, economic, and psychological impacts

Tsunamis produce much more solid waste and disaster garbage, which the authorities can handle. In 2004, disposing and recycling wastes were critically essential priorities. Toxic materials, such as asbestos, oil fuel, and other industrial raw materials and chemicals, are mixed with ordinary garbage. Contamination of soil and water was another serious issue. Water bodies such as rivers, wells, and inland lakes were subject to salination. A further effect is a change in the soil fertility of agricultural lands. In many cases, sewage infiltrated the water supply system.

The estimated cost of the damage from the 2004 tsunami was just under $10 billion [25]. Fishing and tourism are the two major sectors affected by the tsunami. Damage to infrastructure was also a significant factor. Many of the fishing boats were destroyed.

We cannot underestimate the psychological trauma. A big effect was on religiously conservative Islamic society in Western Indonesia. "In Banda Aceh, Indonesia, the hardest-hit area in the world's most populous Muslim country, imams blamed the Dec. 26 tsunami on lay Muslims who were shirking their daily prayers and following a materialistic lifestyle. Others said Allah was angry that Muslims were killing Muslims in ongoing civil strife." [26].

Tsunami forecasting: scope and limits

There were no tsunami warning systems in the Indian Ocean to detect tsunamis or warn the general population around the ocean. Tsunami detection is not easy because while a tsunami is in deep water, it has little height, and a reasonable network of sensors is needed to detect it.

Ocean researchers cannot make specific predictions about when and where the following tsunami will strike. However, the tsunami warning centers may know which seismic tremors will likely produce tsunamis and can issue messages. The warning system has improved in the last twenty years by adding new sensors and using better algorithms. Tsunami caution capabilities have become drastically superior since the 2004 Indian Sea "tidal wave". According to a report from the National Oceanic and Atmospheric Administration (NOAA), researchers are working to assist progress caution center operations and assist communities in arranging to reply [27]. More recent tsunamis (2011: Tohoku and 2018: Palu), have shaped how tsunami risk is perceived and acted upon. Probabilistic tsunami hazard and risk analyses (PTHA and PTRA) have been developed and have proved useful [28]. PTHA aims to estimate the probability of a tsunami intensity measure exceeding a given threshold in a predefined time interval.

Prediction and control of tsunamis [29]

- To build a breakwater, water gate, and plant tsunami control forest to prevent or lower the shock wave.
- To build a higher breakwater in the ports and city roads joint. And to prevent water from sliding through the water gate.
- To put objects that easily cause pollution in concrete walls with intake gates and outlets and to prevent walls from collapsing from high hydraulic pressure.
- To build shelters away from the coast in an uphill place and to build escapes away from the river to make the public easily flee.
- To build some tall and solid buildings as shelters downtown.
- To install billboards, including fleeing paths and emergency measures in crowded places.
- To build a broadcasting system along the coast to issue tsunami warnings and direct how to evacuate and flee.
- Residents living along the coast have to take evacuation drills regularly. And to put the knowledge of precautions against disaster into pedagogical materials.
- To educate the public common sense about tsunamis. For example, if an earthquake happened or quickly ebbed, the public should move to place uphill and inland.

4.5   Lessons Learned and Looking Forward

The climate system is a prototype of complex systems containing positive and negative feedback loops. Positive climate feedback typically means that some initial change in the climate causes some secondary shift that increases the effects of the initial change, thus magnifying the initial effects. The textbook example of the negative feedback mechanism starts with the increase in temperature and the amount of cloud cover. The increased cloud thickness or amount could reduce incoming solar radiation and limit warming.

Most climate scientists agree that humans are causing global warming and climate change. Scientific evidence shows that human activities (primarily burning fossil fuels) have warmed Earth's surface and ocean basins, which continue to impact Earth's climate. Like it or not, this is our current reality. We have to accept that the predictability of the climate system is limited by the fact that the atmosphere's behavior shows chaotic dynamics. Technically, its evolution is sensitive to small changes in the initial condition.

It looks clear that wildfires are out of control, as recent news (2023 November) state [30] from Brazil's wildlife-rich Pantanal wetlands: "Experts say the fires are mainly caused by human activity, especially burning land to clear it for farming. Climate conditions have only worsened things. ... even when animals survive the flames, they risk starvation." Improving the efficiency of the methods for predicting and controlling tsunamis is possible. To be very simple, it needs better technologies and more human attention.

As we saw how successfully technological and biological systems adopted control loops, finding successful strategies for socio-ecological systems, including our whole climate, is a big challenge.

Next, we will discuss the scope and limits of applying feedback control in economic systems.

References

1. More than half of world's large lakes are drying up, study finds; voanews.com May 23, 2023
2. Global forest watch. https://www.globalforestwatch.org/
3. Stocker TF, et al. Physical climate processes and feedbacks. IPCC report. Chapter 7
4. Forster P, Storelvmo T, Armour K, Collins W, Dufresne J-L, Frame D, Lunt DJ, Mauritsen T, Palmer MD, Watanabe M, Wild M, Zhang H (2021) The earth's energy budget, climate feedbacks, and climate sensitivity. In: Masson-Delmotte V, Zhai P, Pirani A, Connors SL, Péan C, Berger S, Caud N, Chen Y, Goldfarb L, Gomis MI, Huang M, Leitzell K, Lonnoy E, Matthews JBR, Maycock TK, Waterfield T, Yelekçi O, Yu R, Zhou B (eds) Climate change 2021: the physical science basis. contribution of working group i to the sixth assessment report of the intergovernmental panel on climate change. Cambridge University Press, Cambridge, United Kingdom and New York, NY, USA, pp 923–1054
5. Plumer B, Shao E (2023) Heat records are broken around the globe as earth warms, fast. New York Times, July 6, 2023

6. Al Gore (2006) An inconvenient truth: the planetary emergency of global warming and what we can do about it. Rodale Books

7. Nunez Ch. Climate 101: causes and effects. 209. https://www.nationalgeographic.com/environment/article/global-warming-real

8. Herring D (2020) Are there positive benefits from global warming? Climate.gov published October 29, 2020

9. Hood L. An inconvenient truth about an inconvenient truth. https://theconversation.com/an-inconvenient-truth-about-an-inconvenient-truth-81799

10. Tyson A, Funk C, and Kennedy B (2023) What the data says about Americans' views of climate change, https://www.pewresearch.org/short-reads/2023/04/18/for-earth-day-key-facts-about-americans-views-of-climate-change-and-renewable-energy/

11. Herein M, Tél T, Hasypre T (2023) Where are the coexisting parallel climates? Large ensemble climate projections from the point of view of chaos theory

12. Falkenberg M, Galeazzi A, Torricelli M, et al (2022) Growing polarization around climate change on social media. Nat Clim Chang 12:1114–1121

13. Cho R (2022) How do we deal with the polarization around climate change? https://news.climate.columbia.edu/2022/09/23/how-do-we-deal-with-the-polarization-around-climate-change/

14. Camps JM. An inconvenient truth for Al Gore: Greta Thunberg's effectiveness. https://croniquessubsidiaries.org/2020/07/25/an-inconvenient-truth-for-al-gore-greta-thunbergs-effectiveness/

15. McGowan J. Greta Thunberg's stand with Gaza is a problem for the climate change movement, Forbes Oct 25, 2023, https://www.forbes.com/sites/jonmcgowan/2023/10/25/greta-thunbergs-stand-with-gaza-is-a-problem-for-the-climate-change-movement/?sh=8a52d84375cc

16. Dawldoff N (2009) The civil heretic. New York Times Mag, March 25, 2009. https://www.nytimes.com/2009/03/29/magazine/29Dyson-t.html?sq=Freeman

17. Top 12 largest wildfires in history. https://earth.org/largest-wildfires-in-history

18. European forest fire information system. https://effis.jrc.ec.europa.eu/about-effis/brief-history

19. 3 reasons wildfires are getting more dangerous—and 3 ways to make things better. The Wilderness Society, May 21, 2019

20. Rosenthal A, Stover E, Haar RJ (2021) Health and social impacts of California wildfires and the deficiencies in current recovery resources: an exploratory qualitative study of systems-level issues. PLoS One 16(3):e0248617

21. Wolters C (2019) Here's how wildfires get started—and how to stop them. https://www.nationalgeographic.com/environment/article/wildfires

22. https://www.dw.com/en/wildfires-are-on-the-rise-in-spain-due-to-heat-and-drought/video-66256246

23. Charnley S, Spies TA, Barros AMG, White EM, Olsen KA (2017) Diversity in forest management to reduce wildfire losses: implications for resilience. Ecol Soc 22(1):22

24. Tsunami: mechanics of a Tsunami wave. https://www.maine.gov/mema/maine-prepares/preparedness-library/tsunami-wave-mechanics

25. Jalan V (2022) A study of the 2004 Indonesian Tsunami: the effects on GDP growth and tourism post-disaster. Claremont Colleges Scholarship Claremont

26. Broadway B (2005) Divining a reason for devastation followers of various faiths differ on natural, supernatural explanations for Tsunamis. Washington Post, January 8, 2005. https://www.washingtonpost.com/wp-dyn/articles/A57758-2005Jan7.html

27. Tsunamis. National oceanic and atmospheric administration. Education Home

28. Behrens J, et al (2021) Probabilistic Tsunami hazard and risk analysis: a review of research gaps. Front Earth Sci, 29 April 2021 Sec. Geohazards and Georisks, Volume 9 - 2021

29. Centel Aater Bureau, Seismological Center, https://scweb.cwb.gov.tw/en-us/guidance/protection/220

30. "Out of control" wildfires are ravaging Brazil's wildlife-rich Pantanal wetlands

Chapter 5
From Laissez-Faire to Greenspan: Feedback Control in Economic Systems

Abstract This chapter analyzes a fundamental question: Should economics be controlled or not? Is the "invisible hand" and the self-regulation of free market capitalism the best possible mechanism, or does the economy need governmental intervention? Minsky's hypothesis suggests that stability implies instability. It is an observational fact that the economy shows a cyclic pattern: business cycles with different frequencies are very general. Uncompensated positive feedback is found to be a general mechanism leading to extreme events, among other hyperinflation.

5.1 The Simultaneous Emergence of Feedback in Technological and Economic Systems

The Reader remembers that Otto Mayr, the famous historian of technology, analyzed the role of feedback systems in technology. Very importantly, he also investigated how the feedback system concept simultaneously emerged in technology and economic thought in 18th-century Britain [1]. Adam Smith (1723–1790), occasionally called the "Father of Capitalism" and "Father of Economics," was the pioneer of the theory of *free-market* and economic liberalism. The central concept in his celebrated book [2] is the *self-regulating* mechanism.

Adam Smith's Inquiry into the Nature and Causes of the Wealth of Nations, first published in 1776, is a book that gave the classical formulation of economic liberalism. It synthesized the ideas of some earlier British and French thinkers into a consistent economic system. The concepts of feedback and self-regulating mechanisms were fully developed in his work. However, as often, the predecessors deserved credit. Smith, who lived in Glasgow, traveled to France as a tutor of a Scottish duke and was exposed by the ideas of a group of French intellectuals, the Physiocrats. They argued for the *Laissez-faire economy*, which opposed any government intervention in business affairs [3]. It is a vital principle of any *free-market* capitalism.

- Laissez-faire is an economic philosophy of free-market capitalism that opposes government intervention.
- The French Physiocrats developed the theory during the 18th century.

© The Author(s), under exclusive license to Springer Nature Switzerland AG 2024
P. Érdi, *Feedback*,
https://doi.org/10.1007/978-3-031-62439-1_5

- It advocates that economic success is inhibited when governments are involved in business and markets.
- Free-market economists built on the ideas of laissez-faire as a path to economic prosperity, though detractors have criticized it for promoting inequality.
- Critics argue that markets need a certain degree of government regulation and involvement.

Standing on the shoulders of the Physiocrats, Adam Smith suggested a mechanism for the distribution of goods by adopting the concept of *invisible hands*. The statement is that the self-interest of the individuals implies the most efficient resource allocation for the whole society, and the argument is that the direct interaction among producers and consumers in the marketplace ensures a balance between demand and supply. The law of supply and demand says that a free market will move toward an equilibrium quantity and price where the plots of supply and demand intersect. Too little supply increases, and too abundant decreases the price.

A laissez-faire economy assumes that the laws of supply and demand operate efficiently. If any social variable, say the balance of trade or the number of available workers, deviates from its equilibrium value, it implies a compensatory answer (as we remember to the Le Chatelier principle mentioned in Sect. 3.2.2.1) due to the self-interest of the members of the general public.

It is remarkable to see how the intellectual connection developed between technology and economics as the concept of self-regulation emerged. When Adams Smith worked on the Wealth of Nations and employed the concept of feedback in an abstract form, some technological feedback mechanisms reached practical significance. We can understand how Smith came up with his social feedback systems. At some point in his past, Smith encountered one of the many existing technological feedback systems, maybe self-regulating furnaces. He thought about it for a while and understood how it worked. This belief or idea may have become more robust due to additional experiences, such as using different devices that give feedback. Adam Smith learned that systems generally can be self-regulating if their internal processes' cause-and-effect relationships are arranged in a closed loop.

Self-governing stands obviously in contrast to the traditional, generally accepted schemes. Before the appearance of the laissez-faire doctrine, a state's economic and political system was hierarchical and controlled by a central authority. Similarly, many popular mechanical systems, automatons, music boxes, and planetary clocks of the Renaissance and Baroque periods were functioning by a rigid, predetermined program. The simultaneous emergence of the feedback system in technology and economic thought in 18th-century Britain departed from these traditional schemes generally accepted in Europe until then and remained in favor of the Continent a good deal longer. Similar developments can be observed in other regions. For example, in the theory of government, the separation of powers has replaced autocracy and absolutism.

Self-regulating systems started to replace externally controlled systems and could maintain or reestablish their equilibrium without external instructions but by adopting purely internal processes. The profound changes in technology and economics reflected the emergence of a new worldview: The Enlightenment arrived.

5.2 From Feedback Loops to Business Cycles

5.2.1 Negative Feedback and Stability

Warren Buffett is known for being a *contrarian investor* [4]. Contrarian investing is an intentional strategy against actual market trends: such people sell stocks when others are buying and buy when most investors sell. This is a negative feedback strategy, which helps reduce market volatility by pushing systems towards equilibrium. Being a contrarian can be rewarding; most of us like to follow our peers. Markets are subject to herding behavior induced by greed and fear, making markets from time to time over- and under-priced.

Negative feedback stabilizes the economy, leading to a stable market share and price equilibrium. Common sense and traditional theory state that equilibrium is the best possible outcome when the market forces are balanced. It results in the most efficient use and allocation of the available resources. As the price decreases, we consumers buy more and more goods, leading to another equilibrium state. The concept of the law of *diminishing returns* prescribes that the principle of "the more, the better" is invalid. Even the phyisocrat Turgot (1727–1781) noticed that the relationship between the result of the production of crops and the invested effort is not linear but goes to saturation. The phenomenon was found in farming, manufacturing, and service as a mechanism to reach a balance. "If a factory employs workers to manufacture its products, at some point, the company will operate at an optimal level; with all other production factors constant, adding additional workers beyond this optimal level will result in less efficient operations [5].

Decreasing return is not the only game in the world of economics. Since people like to imitate the behaviors of others, they start to buy the same things as their peers, use positive feedback mechanisms, or *increasing returns*, to s to scale use the terminology of economist [6]. This mechanism amplifies small economic changes initially, and predictable, stable markets are no longer guaranteed. Even multiple equilibrium points can be implemented by positive feedback mechanisms. With positive feedback, increasing market share results in increasing returns, which might imply a very unstable, volatile market.

5.2.2 Positive Feedback: Growth or Decline?

Positive feedback is identified as a mechanism based on herd mentality, which may generate irrational exuberance when buying or selling assets, resulting in financial bubbles. A bubble is never ever-lasting. A stock market crash is an abrupt fall of stock prices in a significant stock market region. There have been several big crashes in the last few centuries. The interplay of irrational expectations and specific economic factors generates boom-and-bust financial dynamics. Historical analysis shows that positive feedback leading to a super-exponential increase between the realized growth rate and people's expected growth rate also led to hyperinflation. The market's self-regulatory effect seems insufficient to avoid significant volatility. Once again, such crashes emerge due to the interaction between inherent economic factors and human actions.

The Tulip Bulb Mania

The Tulip Bulb Mania emerged in Holland at around 1635. Tulip bulbs were first bought for their aesthetic value, but as their prices increased, they became subject to buying and selling. People bought them to make a (significant) profit. There was a month when its value increased twenty-fold. At a certain point, the Dutch government attempted to control the mania. After its regulatory actions, some informed speculators realized that the price could not become more inflated and started to sell bulbs. Other people soon noticed that the demand for tulips could not be maintained. Their attitude propagated rapidly among people interested in the business, and soon panic, a *social collective phenomenon* emerged. During six weeks, there was a 90% reduction in its price.

The late 1990s had the dot-com bubble. Investors started to pour money into internet startups, hoping they would be profitable. New companies rose without having any reasonable trade plans, no items or administrations prepared to bring to marketing, and in numerous cases, nothing more than a title (as a rule, something tech-sounding with ".com" or ".net" as a postfix). Despite missing in vision and scope, these companies pulled in millions of speculation dollars and saw sky-high stock costs. However, the bubble was not sustainable. We will return to the history and predictability of the stock market crash.

John Maynard Keynes (1883–1946) argued [7] that positive feedback might generate a negative spiral, too. If, for some reason, the consumer's confidence is reduced, spending is decreased. The following grain in the chain is the decrease in production, followed by an increase in unemployment. The next cycle starts with a further reduction of consumer confidence and so on...eventually leading to recession. Keynesian theory does not see the market as being able to restore itself naturally. Therefore, Keynes suggested the necessity of economic intervention by the governments to stop the negative spiral of the Great Depression: the government should inject money into the system to stop reducing unemployment with all of its further negative consequences. It would increase consumption, the critical factor of any economic recovery. Keynes's suggestions were in strong contrast with the laissez-faire principle. It is well known that Milton Friedman (1912–2006) initiated the anti-Keynesian counterrevolution. He did not object to governmental intervention but suggested a steady, small expansion of the money supply compared to rapid and significant changes.

Feedback is the transmission and return of information about the "state" of a system, i.e., about the amount of "stuff" accumulated in each of a system's stocks over time. This information travels throughout a system and eventually returns to the flows that fill or drain the stocks, thus closing the system's feedback loops. Generally speaking, the information being transmitted via feedback loops is used by agents to make decisions that alter a system's flows and cause the system to adapt its behavior over time.

Positive feedback represents self-reinforcing processes and is generally responsible for systems' growth or decline. Economic growth trends, multiplier processes, accelerator relationships, wage-price spirals, speculative bubbles, bandwagon effects, increasing returns, path-dependent processes, and anything that can be described as a vicious or virtuous circle can be represented with positive feedback loops. Financial panics and market crashes are examples of positive feedback in markets that head in a negative direction. Bubbles are positive feedback loops that instead send prices higher. From a psychological perspective, reinforcing feedback is also a self-fulfilling prophecy. It is a form of reinforcing feedback that results from the interaction between the economy and agents' perceptions and expectations.

Historically, a self-reinforcing mechanism was identified as a driving force behind the success story of the Japanese car industry. Lower costs and higher quality provided advantages to other countries. It started around the early 1970s when some Japanese companies began to sell a relatively large number of small cars in the United States. Detroit could not react rapidly, so cars made in Japan gained market share. Their engineers and workers in production became more and more experienced, and they produced cars with even better quality and lower prices. The sales activity has also increased; Detroit's answer was slow, so the positive feedback loop helped Japanese companies sustain their selective advantage for small cars in the US [8].

Negative loops, on the other hand, represent goal-seeking processes and many types of purposeful behavior. They can either stabilize systems or cause them to oscillate when their corrective action is delayed (by stocks). As such, negative loops are responsible for such phenomena as the "invisible hand" equilibrating processes of well-functioning markets and the unstable behavior of macroeconomic cycles.

5.2.3 Nonequilibrium Economics: Route to Complexity

Even before Adam Smith, economists noted that macro variables in the economy, such as market prices and quantities of goods produced and consumed, are aggregates of micro variables characterizing individual behavior, and individual behavior, in turn, reacts to these macro variables. There is a recursive micro-macro loop. In equilibrium, the players have no reason to change their strategic behavior. But if this is the case, there is no incentive to create new products, establish new institutions and organizations, or think of new strategies.

The real-world economy is now seen as a nonequilibrium system [9–11].

Dynamics

It would be difficult to deny that the economy changes over time; it is a dynamic system. In Sect. 1.2.3, it was briefly mentioned how system dynamics based on *stock* and *flow* variables emerged as a helpful simulation methodology in social sciences. Stock (a word with multiple meanings!) is an accumulated quantity, say, the actual balance in your bank account. The velocity of the change of stock variable is a flow. You can think of the economy as a set of stocks of flows connected by complicated loops. Say, if the stock of employment is being reduced, there might be other changes; interest rate reduction could lead to the increase of stock of money, and in the next step, new investments emerge, and they would lead to increased employment. Feedback loops drive a dynamic economy. Figure 5.1c shows the basic qualitative behavior in dynamic systems, excluding chaos.

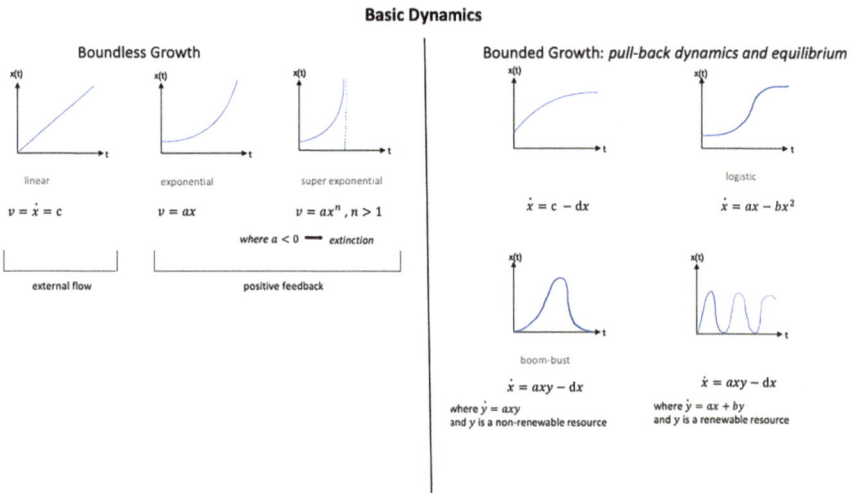

Fig. 5.1 Basic dynamics. The left side of the panel shows the three basic subclasses of boundless growth: the graphs of linear, exponential, and superexponential dynamics and the models leading to such behaviors. The upper two graphs demonstrate two ways to converge to equilibrium: without (left) and with (right) inflection point. The lower two plots show boom-and-bust and periodic behavior, respectively. These are simple models of phenomena when the resources are non-renewable (left) or renewable (right)

There are more complicated situations. We can imagine that the velocity of the change explicitly depends on the past. In finance, Friedman famously argued about the effect of time lag: There is much evidence that monetary changes have their effect only after a considerable lag over a long period and that the lag is variable [13]. Dynamics and control of a financial system with time-delayed feedback can be studied using sophisticated mathematical models [14].

5.2.4   Business Cycles and Chaos

It is an observational fact that an economy shows expansion (growth) and contraction (recession). Economists like to divide the process into four stages in a cyclical pattern: expansion, peak, contraction, and trough. The cycle period varies, so explaining the emergence of oscillations by the mathematical model adopted in physics and related areas is difficult. The cycle is sitting on a trend; an idealized cycle is shown in Fig. 5.2.

Real-world economic time series are often neither random nor regular. As we know from the theory of complex systems [12], complexity lives between purely deterministic and purely random systems. Markets are complex and chaotic systems, and their behavior has both systemic and random components. As classical theory suggests, stock markets are not totally unpredictable but have severe limits.

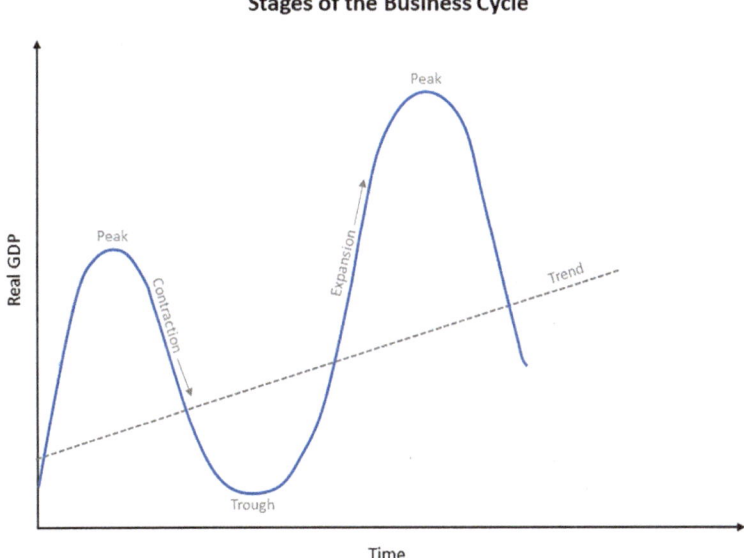

Fig. 5.2 The four stages of a business cycle are shown: expansion, peak, contraction, and trough. The cycle is sitting on a trend. From [15]

A financial market is highly complex, and the only real prediction that we can make is that it is unpredictable. The unpredictability of the financial market is due to the complexity and randomness of many events. By adopting chaos theory, one can try to catch the structure of unpredictability. Chaos theory contributed to understanding and forecasting the behavior of financial markets. Chaos theory appeared during the transition of finance to the use of big data. With the development of new algorithms, some pseudorandom events turned out to be deterministic but irregular (i.e., chaotic).

Ideally, the financial system should show security and stability. It is a fundamental assumption to have stable economic and appropriate social development. Financial globalization and liberalization increased the complexity of the financial system. Mathematically, these systems are nonlinear and might show financial chaos (such as financial market turbulence and crisis).

So, how do we live with chaotic economics? First, technically, it is possible to design algorithms to tame chaotic patterns. An intermittent feedback controller was designed for the chaotic financial system. Adjusting the controller parameters allows the financial system to be controlled from chaotic to periodic temporal patterns [16]. Second, we can still make valid stock market forecasts while markets are chaotic systems with complex dynamics. Using these forecasts generated by cutting-edge predictive algorithms and a careful risk management strategy may give a trader a significant competitive advantage.

We cannot avoid a difficult question (and some more difficult answers) discussed in the next section.

5.3   Is the Financial System Inherently Unstable?

The Tulip Bulb Mania was just the first example of repeating events in the history of the financial crises. To give a few selected examples:

- Tulip mania Bubble (1637)
- South Sea Bubble (1720)
- Wall Street Crash (1929)
- Black Monday (19 Oct 1987)
- Japanese asset price bubble (1991)
- Russian financial crisis (17 Aug 1998)
- Dot-com bubble (2000)
- Great Recession (2007–2008)
- cryptocurrency crash (2018)

From Mania via Panic to Crash. Typical patterns form the anatomy of bubbles and crises: 1. When people and companies live in a stable economic environment, some perceive small, favorable changes in the circumstances and are ready to take risks. 2. *Expected* profit is increased by a positive feedback mechanism leading to social euphoria. 3. The situation attracts even novices; these first investors form a manic mass. 4. Some insiders learn that the (maybe super-)exponential growth cannot be sustained and start to sell shares. 5. The outsiders suck, panic emerges, and they try to sell their shares with a decreased value.

The bottom line is that the classical economic theory intentionally neglects the psychology of the participants and assumes that the participants are rational and have complete information. However, manias, panics, and crashes are an inherent feature of markets, as among others, the many editions of the bestseller *Manias, Panics, and Crashes: A History of Financial Crises* by Charles Kindlerberger (1910–2003) states [17]. Along this line of thinking, more and more economists accept the **The Financial Instability Hypothesis** suggested by Hyman Minsky (1919–1996). The main point is that most financial bubbles and crashes are *internally generated*. The classical theory explains bubbles and crashes caused by external disturbances, such as wars and diseases.

We cannot deny that the COVID-19 epidemic is an external disturbance to the world economy. It generated shock waves worldwide and triggered the most significant global economic crisis in over a century. The crisis dramatically increased inequality both within and across countries. Still, the main mechanisms of financial bubbles and crashes are the inherent features of the economy.

The primary statement of the financial instability hypothesis is somewhat paradoxical: stability implies instability [18]. Longer-term stability gives the feeling that the economic environment is safe. People with risk-taking character make more investments. (Some people sold their houses in Holland to buy tulip bulbs.) That risk-taking creates financial bubbles that eventually result in panic and crisis when the economy collapses. Minsky identified three stages of lending he labeled as hedge, speculative, and Ponzi.

Initially, lenders and borrowers are generally cautious during the hedge stage. Both borrowers and lenders take reasonable risks. In the next stage, called speculative, the primary assumption of the participants is that an even more significant risk might be sensible. In search of higher returns, borrowers are ready to take on more substantial amounts of debt since lenders reduce the standard of credit granting based on the hope that asset prices will continue to rise. Borrowers typically can cover the interest on their loans but may have difficulty repaying the principal. What next? A very influential economist, Kenneth Galbraith (1908–2006), wrote an easily readable small book titled "A short history of financial euphoria." [19]. The book explains why and how the interplay of irrational expectations and specific economic factors generated boom and bust financial dynamics over several decades. So, the final, so-called Ponzi stage emerges. Since lenders and borrowers have forgotten the lessons of the prior crisis, the participants are overconfident and convinced that asset prices will continue to rise. Too much risk-taking and debt produced a financial environment similar to a house of cards. So, the increase in the asset price cannot be sustained, and the market collapses accompanied by panic very rapidly at the "Minsky Moment."

Mainstream financial experts did not accept Minsky's hypothesis while he was alive. After he died in 1996, the dot-com bubble and the Great Recession of 2008 justified his explanation. His theory is now generally accepted as a primary explanation for the boom-and-bust cycles in the economy.

What Minsky's Hypothesis can and can't explain

The hypothesis can be considered as a conceptual model. Figure 5.3 summarizes the stages of the cycle and the transition among them.

It helps people orient where the word is in the cycle. However, the duration of the stages, such as euphoria or panic, varies. It is difficult to predict precisely when a market euphoria or stock market bubble reaches its maximum. (It would be interesting to do some statistical analysis of the length of these stages.) Pandemics, wars, and other geopolitical events are external effects that may have significant influences. So, we can not have reliable predictions about when the economy will transition from one part of the cycle to the next.

However, knowing generally where we are within the cycle can illuminate great techniques for financial specialists and commerce proprietors. As the economy and markets move from boom to euphoria, it is fundamental to have a healthy margin of security within the frame of cash and high-quality bonds. Savvy businesses will increment their cash to shore up liquidity and stand up to the allurement to require more obligation. At that point, when the profit-taking and freeze happen, they can redeploy their security edge into bargain-priced hazard resources.

It is impossible to accurately predict the timing of the market's tops and bottoms. However, we may know roughly where the economy is in the cycle, which may help with investment decisions.

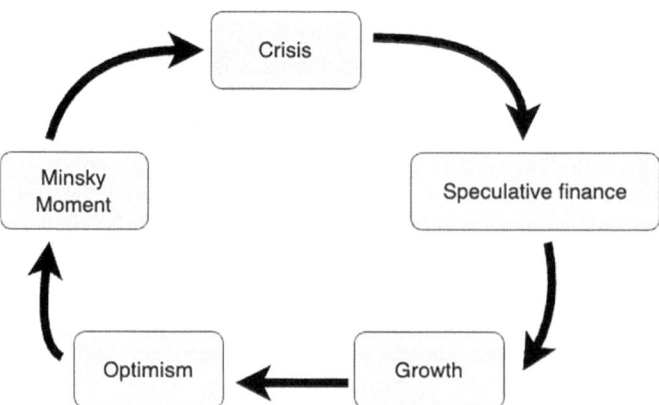

Fig. 5.3 Stylized cycle based on the Minsky Hypothesis. based on [20]

5.4 Lack of Self-regulation: From the Doom Loop Theory via Hyperinflation to Greenspan's Error

5.4.1 Doom Loop Theory

Loosely speaking, a *doom loop* is a mechanism that makes bad things even worse. Consider a chain of events, beginning with a negative one, which triggers another, which in turn triggers either a new bad event or worsens a previous one [21]. The Greek debt crisis around 2010 is a textbook example of this theory. A chain of adverse events (technically, not necessarily a loop) followed each other.

- Prior governments had been distorting national financial information.
- The new government disclosed a budget deficit much worse than expected.
- Credit rating agencies reacted to downgrade Greece's government debt to junk status.
- Fear propagates through other eurozone countries.
- Lenders increased interest rates.
- Governments cut spending and raised taxes, slowing the economy down.
- Investors started selling bonds, so local banks became weak.
- Bailout arrives from external financial institutions under the condition of applying austerity measures.

A stylized scheme of doom loop is shown in Fig. 5.4.

Another application of the theory is *urban doom loop*. COVID-19 dramatically changed our life. People prefer working remotely from their home offices. Work from home triggered an office real estate apocalypse, and soon, as now-empty offices were subject to crime, unsheltered homelessness rose. Of note, the challenges of downtown started much before the pandemic.

Fig. 5.4 Stylized Doom loop. Based on [22]

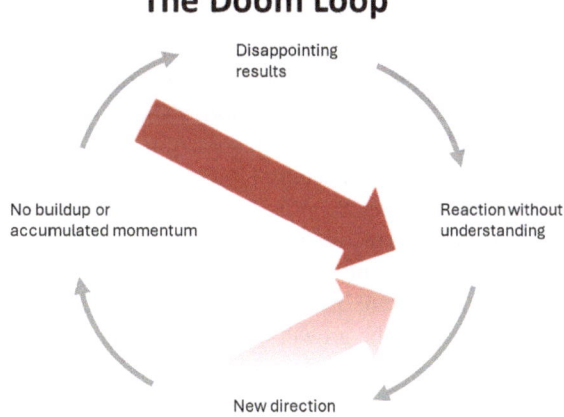

The Doom Loop

Disappointing results

Reaction without understanding

New direction

No buildup or accumulated momentum

I corresponded with my college schoolmate, Ginger (Gyöngyvér), who lives in San Francisco, the textbook example of an urban doom loop.

The decline and eventual death of San Francisco is a fashionable topic and a good opportunity for the press and conservative politicians to illustrate the failure of left-wing liberal politics.

I apologize for being biased. I am emotionally attached to this city, which is often toxic, chaotic, and self-destructive but beautiful, never dull, and intellectually inspiring. Throughout its history, the city has been predicted many times to fall apart and die, such as after the 1906 earthquake, after the 1967 hippie invasion (Summer of Love), during and after the AIDS crisis of the 1980s, and after the dot-com bubble burst around 2000. The news of the city's death has always been premature. I hope and believe it is happening now.

That does not mean the situation is not severe. Many of the downtown office buildings are empty, partly because people in certain professions have discovered the opportunities and benefits of working from home, partly because companies and businesses have moved to cheaper locations, partly because of the drug and homeless crisis on nearby streets and the tents and trash blocking the roads. The city's leadership is pulling apart, bickering among themselves instead of compromising, and common sense solutions often fail because of court-backed ultra-leftist policies. The causes and possible solutions to the drug and homeless crisis are very complicated and far beyond my ability and the scope of this short write-up.

However, there are voluntary, severe grassroots movements that will hopefully improve the situation. The problems are serious, but I don't think they are insurmountable.

Doom loop theory is not the only game in town. The business community calls the analogy a *flywheel*. This is a mechanism to amplify initial small advantages by a positive feedback loop [22] and an explanatory theory of why some companies make the leap, and others don't.

5.4.2 Hyperinflation

5.4.2.1 General Mechanisms

Hyperinflation is an extreme and rapid increase in the general price level of goods and services within an economy. It is characterized by inflation rates exceeding 50% per month, often leading to a complete breakdown of the economy and a loss of confidence in the national currency. This rare phenomenon can devastate a nation's economic stability and social fabric. The mechanism of hyperinflation is complex and typically involves a combination of factors. Here's a list of steps and an overview of how hyperinflation occurs:

Excessive Money Supply: Economists generally assume that hyperinflation starts with a significant increase in the money supply. This can happen for various reasons, including a government's decision to print more money to finance budget deficits, war expenditures, or bail out failing financial institutions. The increase in the money supply outpaces the growth of the real economy, creating an excess of currency in circulation.

Loss of Confidence: As the money supply rapidly expands, confidence in the national currency erodes. People start to lose faith in their currency's stability, leading to a vicious cycle. The more people believe the currency is losing value, the faster they spend it, further accelerating inflation.

Wage-Price Spiral: Prices of goods and services begin to rise rapidly due to the excess money in circulation. In response, workers demand higher wages to maintain their purchasing power. Employers raise wages, passing on the increased labor costs to consumers through higher prices for goods and services. This creates a self-perpetuating cycle in which rising wages and prices fuel each other.

Speculation: People and businesses may start speculating in the foreign exchange or asset markets to protect their wealth from the depreciating domestic currency. This can exacerbate the problem as they exchange their local currency for more stable assets or foreign currencies, further devaluing the domestic currency.

Shortages and Hoarding: As hyperinflation accelerates, shortages of essential goods can occur. People may begin hoarding goods, fearing that prices will rise even further. These shortages can lead to even higher prices, creating a positive **feedback loop**.

Monetary Policy Failures: In many cases of hyperinflation, the central bank loses its independence and becomes subject to political pressure, leading to poor monetary policy decisions. These policies can include excessive money printing, artificially low interest rates, and a lack of fiscal discipline.

Loss of Trust in Government: Hyperinflation often reflects broader economic and political *instability*. Citizens lose trust in the government's ability to manage the economy and maintain price stability. This can lead to civil unrest and regime change, further destabilizing the economy (see more in Sect. 6.2.3).

External Shocks: In some cases, external factors like war, sanctions, or natural disasters can trigger or exacerbate hyperinflation by disrupting the normal functioning of the economy.

The theory of Didier Sornette, a leading expert in predicting extreme events, suggests that the transition from a relatively normal to hyperinflation state happens without any external triggering event [23] and by using positive feedback loop (public loss of confidence in the stability of currency -> hoarding goods -> shortage -> price increase -> further loss of confidence).

In summary, hyperinflation is a complex and **self-reinforcing** economic generated by a positive feedback loop producing even super-exponential increase without having any damping term. A textbook example of a doom loop, *uncompensated positive feedback* seems to be a general mechanism behind phenomena like earthquakes, epileptic seizures, stock market crashes, hyperinflation, and political instabilities.

5.4.2.2 The Hungarian Crises in 1923 and 1946

Positive feedback significantly influenced the periods of Hungarian hyperinflation in 1923 and 1946 [24].

After World War I, Hungary was required to pay war reparations to the victorious Allies. These payments placed a significant strain on the Hungarian economy. The positive feedback loop was that the higher the inflation rate, the more Hungary printed money to make these payments. As in the next step, the currency was devalued, and they needed to print more money to make the duplicate payments. Hyperinflation was close. There was also a loss of confidence. With increasing inflation, people lost confidence in the Hungarian Pengö, leading to a vicious currency depreciation cycle. People expected their money to lose value rapidly, so they spent it as quickly as possible, further driving up prices. The Hungarian government's response to the hyperinflation was counterproductive. They printed even more money to cover budget deficits, exacerbating the problem. This lack of fiscal discipline and economic mismanagement only made matters worse.

The situation was not better in 1946. Hungary faced a devastating post-World War II situation. The country was in ruins, and the government had accumulated huge debts. The feedback loop began when the government resorted to printing massive amounts of money to fund reconstruction efforts and pay off debts. The country experienced severe food shortages, leading to a black market where people traded goods for foreign currencies. This further eroded trust in the local currency, leading to its devaluation. At a certain point in July 1946, prices doubled every 15 hours.

The introduction of the new forint currency was a crucial step in ending the hyperinflation. The old pengö currency was replaced with the new forint (HUF), with the exchange rate of 400,000,000,000,000,000,000,000,000,000,000 (four hundred quadrillion) pengö for 1 forint. This massive devaluation essentially wiped out the old currency, introduced a new, more stable one, and allowed Hungary to start rebuilding its economy.

5.4.3   Greenspan and Beyond

I am inclined to think that one of the most important statements of this century came from Alan Greenspan, who served as the Chairman of the Federal Reserve for eighteen years. He acknowledged he made mistakes in his approach to financial regulation by assuming that banks, operating in their self-interest, would do what was necessary to protect their shareholders and equity. Greenspan's admission came amid the Great Recession, which revealed severe flaws in the financial system and led to widespread economic turn-moil [25].

This acknowledgment was significant because Greenspan was a loyal advocate of free markets and minimal regulation. His concession highlighted the need for reevaluating the regulatory framework governing financial institutions and contributed to ongoing discussions about the role of government oversight in the economy.

As an opinion in the *New York Times* wrote [26]:

> Mr. Greenspan is right on one thing. The "whole intellectual edifice" collapsed. But he is wrong to blame it solely on the wrong inputs. It is too bad that Mr. Greenspan never appreciated the work of Hyman Minsky, who understood that stability is destabilizing and that there will come times when the very calmness of markets and lack of apparent risk causes investors to take ever greater and greater risks. **What was missing was a regulator** [my boldface. PE] who understood markets rather than worshipped them.

5.5 Open and Closed Loop Recycling: Connecting Technology and Economy

Open and closed-loop recycling are two distinct approaches that connect technology and the economy differently. These approaches aim to reduce waste, conserve resources, and promote sustainability, but their processes and economic implications differ.

Open loop recycling, also known as downcycling, involves transforming a product or material into a different product with reduced quality or functionality. This process is common in recycling paper, glass, and certain plastics. The original material or product does not return to its initial state or use but is instead *repurposed* for a different application. For example, recycled glass may be used for road construction instead of being turned back into glass containers. It can be economically beneficial as it often requires less energy and resources compared to producing new materials from scratch. However, it may not create as much economic value as closed-loop recycling due to the loss of product quality.

Advanced Sorting Technologies play a crucial role in efficiently sorting and processing various materials. Automated sorting systems equipped with sensors and artificial intelligence can identify and separate different types of materials, enhancing the effectiveness of open-loop recycling processes. Research and development in *material science* and technology lead to the creation of new materials that can be derived from recycled products. This innovation creates opportunities for the use of recycled materials in novel applications.

Closed Loop Recycling, also known as circular economy or closed-loop supply chain, focuses on maintaining the quality and functionality of products or materials as they are recycled and reused. The goal is to create a *continuous loop* where products return to their original state. It is commonly associated with high-value materials, such as electronics, where products are disassembled, and the components are reused or remanufactured to maintain their quality and performance. It can create more value by extending the life of products and reducing the need for new manufacturing. It also promotes job creation and innovation in technologies that support the closed-loop process. It often requires *advanced recycling technologies*, such as chemical and mechanical recycling processes that can break down materials into their original form without significant quality loss. Investment in research and development of these technologies is crucial for closed-loop recycling success.

Closed-loop resource management ideally produces zero waste, as all products at the end of their lifecycle become assimilated by either technical or natural systems to their benefit. In a broader context, zero waste technology covers zero emission and zero water pollution. The goals are ambitious, and we are far from being there [27].

5.6 Lessons Learned and Looking Forward

Keynes developed theories in the early 20th century that the Federal Reserve in the US still uses to manage monetary policy today. Most modern economic theories discuss the contradiction between Keynes and the free-market theory of Milton Friedman. This does sound oversimplified, but at the end of the day, the question is "Should we or should we not?" (control economics). As we finally learned from Greenspan, we should. But how? This is another question.

Classical economics was extended by behavioral economics, which considers the fact that people are not necessarily rational, as is assumed by the classical theory, and are instead biased [28, 29]. Behavioral economics overlaps classical economics and psychology. It is the study of the effect that psychological factors have on the economic decision-making process of individuals.

The legendary Indian economist from Harvard, Amartya Sen introduced ethical concepts [30]. He proposed a social welfare function (SWF) that penalized economic inequality. Sen used income measures and the Gini coefficient (G), a quantitative measure of economic inequality, to formulate his function. G is zero in the case of perfect equality (i.e., everyone has precisely the same level of income), and G is one in the case of extreme inequality (i.e., one person receives all the income, and others receive none). Sen's SWF is defined as a country's average per capita income multiplied by $1 - G$. Sen's suggestion is *normative*: it prescribes how society **should** behave.

We have another approach called *descriptive*. It focuses on how the society behaves. As we saw from the debate after Greenspan's statement, *"What was missing was a regulator"*. Collective wisdom states that in the US, Republican politicians tend to oppose government intervention in the economy. In contrast, Democratic politicians suggest that the government regulate the economy and mainly fund social programs through taxation.

Well, who should regulate? I want to say that it should be the democratic institutions. However, there are growing concerns about the increasing concentration of political power in a corporate and financial elite that has been able to influence the rules that run the economy and that the rules do not represent the interest of the majority. We should remember that social rules differ from rules of nature since they are human constructions. If rules are made primarily by people with power and wealth, there might be a vicious cycle [31], an amplifying mechanism that helps the very rich to buy politicians and even courts. This mechanism implies an upward distribution of income and wealth from the bottom 90 or so percent to the top. However, the vicious cycles are never endless; we may hope that virtuous cycles may emerge at least to compensate for the adverse effects. In Sect. 6.2.3, we will discuss how much social unrest we need. As a professor at a liberal arts college for more than two decades, I should trust that our students will be among the leaders of a new, compensatory social movement.

References

1. Mayr O (1971) Adam Smith and the concept of the feedback system: economic thought and technology in 18th-century Britain. Technol Cult 1–22
2. Smith A (1976) The wealth of nations edited by R. H. Campbell and A. S. Skinner
3. https://www.investopedia.com/terms/l/laissezfaire.asp
4. https://www.investopedia.com/terms/c/contrarian.asp
5. Hayes A (2022) Law of diminishing marginal returns: definition, example, use in economics, https://www.investopedia.com/terms/l/lawofdiminishingmarginalreturn.asp
6. Arthur B (1994) Increasing returns and path dependence in the economy. University of Michigan Press (1994)
7. Keynes JM (1936) The general theory of employment, interest, and money. Palgrave McMillan
8. Arthur B (1989) Competing technologies, increasing returns, and lock-in by historical events. Econ J 99:116–131
9. Kornai J (1972) Anti-equibirium North-Holland Publishing Company (in Hungarian)
10. Berger S (2009) The foundations of non-equilibrium economics: the principle of circular cumulative causation. Routledge
11. Martinas K (2006) Non-equilibrum economics. Interdisc Descr Complex Syst 4(2):63–79
12. Érdi P (2007) Complexity explained. Springer
13. Friedman M (1961) The lag in effect of monetary policy. J Polit Econ 69:447–466
14. Chen W-C (2008) Dynamics and control of a financial system with time-delayed feedbacks. Chaos Solitons Fractals 37:1198–1207
15. Wolla S (2023) All about the business cycle: where do recessions come from? Page One Economics, https://research.stlouisfed.org/publications/page1-econ/2023/03/01/all-about-the-business-cycle-where-do-recessions-come-from
16. Lu X (2020) A financial chaotic system control method based on intermittent controller. Math Problems Eng 2020, Article ID 5810707, https://doi.org/10.1155/2020/5810707
17. Aliber RZ, Kindleberger CP, Cauley RN. Manias, panics, and crashes: a history of financial crises. Palgrave Macmillan, 8th ed. 2023 edition
18. What is a Minsky moment? definition, causes, history, and examples, https://www.investopedia.com/terms/m/minskymoment.asp
19. Galbraith K (1990) A short history of financial Euphoria. Penguin Publishing Group
20. Urillo B (2016) https://www.slideshare.net/IrulloBeatrice/pinskys-financial-instability-hypothesis—
21. McGrath N. Doom loop: definition, causes, and examples. Investopedia
22. Collin J (2001) Good to great: why some companies make the leap and others don't HarperBusiness, 1st edn
23. Sornette D, Takayasu H, Zhou WX (2003) Finite-time singularity signature of hyperinflation finite-time singularity signature of hyperinflation. Physica A 325:492–506
24. Hartwell CA (2019) Short waves in Hungary, 1923 and 1946: persistence, chaos, and (lack of) control. J Econ Behav Organ 163:532–550
25. Galbraith JK (2006) Greenspan's error. In: Unbearable cost. Palgrave Macmillan, London
26. Norris F (2008) Greenspan's Lament. NY Times October 23, 2008, https://archive.nytimes.com/economix.blogs.nytimes.com/2008/10/23/greenspans-lament/
27. Recycling: open-loop versus closed-loop thinking, https://www.e-education.psu.edu/eme807/node/624
28. Simon H (1957) A behavioral model of rational choice. In: Models of man, social and rational: mathematical essays on rational human behavior in a social setting. Wiley, New York
29. Kahneman D (2011) Thinking. Fast and slow, farrar, straus and giroux
30. Sen A (1997) Choice, welfare, and measurement. Harvard University Press, Cambridge
31. Recih R (2020) The system. How rigged it, how we fix it. Knopf Publishing
32. https://www.globalcitizen.org/en/content/does-multiculturalism-work-social-diversity-global/

Chapter 6
From Natural Disasters to Social Riots

Abstract This chapter starts with analyzing whether we should worry about the possibility of existential risk. In the subsection about the complex system approach to political instability, several issues, namely terrorism, social unrest, and migration, are discussed. Finally, a short system-theoretical analysis implies that democracies do better than autocracies.

6.1 Collapsology

6.1.1 Should We Worry About the Probability of Existential Risk?

Collapsology is a new transdisciplinary scientific field to study the possible mechanisms of the collapse of human civilization [1, 2]. Collapsologists and researchers are trying to estimate the probabilities of risks. They discriminate between *global catastrophic risk* and *existential risk*. A global catastrophic risk is an event that could cause severe and widespread harm to human well-being on a global scale. While it may not necessarily lead to the extinction of the human species, it could have devastating consequences for civilization as we know it. An existential risk is a threat that can potentially lead to the **extinction of humanity** or cause irreparable damage to human civilization, resulting in a permanent collapse. Existential risks are events that could fundamentally and **irreversibly** alter the course of human history. Large-scale natural disasters (such as supervolcanic eruptions, asteroid impacts, or extreme climate change), global pandemics, nuclear war, and specific technological risks fall under global catastrophic risks. *Global* atomic war, artificial intelligence surpassing human control, catastrophic misuse of advanced biotechnology, or a global unrecoverable ecosystem collapse are considered existential risks. There are several classes of sources of risks:

(i) Natural existential risk

So far, humanity has survived heat waves, droughts, wildfires, storms, floods, and other climate events. Will we have a catastrophic natural disaster in the remaining

© The Author(s), under exclusive license to Springer Nature Switzerland AG 2024
P. Érdi, *Feedback*,
https://doi.org/10.1007/978-3-031-62439-1_6

part of this century? As is known from the elementary theory of logic, induction, i.e., the method to make implications from the past to the future, never leads to certain conclusions. So, it **may** happen. However, based on historical records and the assumption that the circumstances do not change dramatically, the probability that a human extinction happens in this century due to natural disasters is relatively small [3].

Controlling natural disasters

Our ability to *control natural disasters* would significantly contribute to building preservation, the safety of people, and lifestyle stability. While it is difficult to imagine how we could control earthquakes by preventing buildings and infrastructure from collapsing, the number of injuries and fatalities would be dramatically reduced. In the same spirit, by controlling wildfires, the loss of forests and houses would be much less significant. It would be very favorable for the environment and the economy.

While scientific and technological progress has allowed us to predict and prepare for natural disasters, our ability to control them is still minimal. "Many natural disasters are caused by complex geological or meteorological processes that are difficult to understand and even harder to control." [9]

Nowadays, more concern exists about the *anthropogenic* existential risk, which derives from human activity.

(ii) Nuclear war

In 1945, the United States dropped atomic bombs on the Japanese cities of Hiroshima and Nagasaki. This was the first and, so far, the only use of nuclear weapons in war. Despite the massive buildup of these weapons, they have never been used in war again; however, the risk of them being used is significant and growing.

The efficiency of the atomic bomb roughly implied two types of answers. The first reflects the spirit of the positive feedback: *It works, we should build more efficient bombs!* The second expresses the *deep concern* about the possibility of what we now call existential risk due to nuclear weapons and offers a negative feedback mechanism to avoid the explosion of the development of nuclear weapons. Leo Szilard (1898–1964) famously came up with the idea of the nuclear chain reaction and later initiated the Manhattan Project (led by Rudolf Oppenheimer (1904–1967)) that built the atomic bomb. After Hiroshima and Nagasaki, his sounds led to alarm politicians against the possible development of thermonuclear bombs, a new kind of nuclear weapon that might annihilate mankind.

The theory behind the second-generation nuclear weapon, the thermonuclear weapon, also called hydrogen bombs or H-bombs, was initiated by Enrico Fermi (1901–1964) and Edward Teller (1908–2003). Teller is often labeled as "the father of the hydrogen bomb," and he worked on the project years before 1945. However, after the end of the war, there was no immediate need to have an H-bomb. As the Soviet Union developed its first atomic bomb in 1949, there was a debate about the necessity of building the H-bomb. Oppenheimer and Fermi famously opposed the development of the hydrogen bomb when it was discussed during a 1949–1950 governmental debate. It is also well-known that Teller played a significant role by witnessing Oppenheimer at the McCarthy trials. His testimony contributed to Oppenheimer's security clearance being denied.

Let us now turn back from history to the near future problems. Nuclear-armed states should eliminate their nuclear arsenals before they eliminate us. Nuclear war itself, whatever dangerous it is, would not lead to total human extinction. Places with low population density may survive with higher probability than big cities.

Calculations from the 1980s showed that smoke from fires following a nuclear war could have dramatic climate effects. The term *nuclear winter* was coined. Nuclear winter theory [4] suggests that nuclear weapons generate fire, which induces smoke. It might happen that black smoke could block out sunlight over the whole planet. There are detailed scenarios of how nuclear winter could be produced.

About 12,500 nuclear warheads were worldwide in January 2023 [5]. Almost 90 percent of them belong to the United States and Russia. In 1985, the number was 63,632, so we may believe that the world became safer in this respect. The newly emergent nuclear powers threaten the superpowers, but most likely not with extinction, "only" with severe consequences, including drought and famine.

(iii) Climate change

There are reasonable arguments that, if uncontrolled, climate change could lead to severe consequences. It may induce food and water shortages, leading to geopolitical instability and implying massive intercontinental migration. These series of events might pose a threat to the existence of human civilization. While some people rightfully say we do not have reliable methods to predict the likelihood of complete extinction, mainstream IPCC climate models do not predict human extinction.

How do we cope with the situation we have? International agreements, technological innovations, and policy measures help to address climate change. Several key climate actions were identified [6]: greenhouse gas emissions reductions and carbon removal, clean mobility, renewable energy, optimization, land restoration, and no deforestation. The dynamics of future emissions, the success of appropriate control strategies, and the ability of societies to adapt will play crucial roles in determining the ultimate impact of climate change.

(iv) Man-made pandemics

We know from history that pandemics have caused significant loss of life and societal disruption, but they have not led to extinction. I think it is wise to consider the COVID-19 pandemic as a warning signal. It has highlighted that local and global institutions were ill-prepared to respond to an event of this magnitude. Massive progress in biomedical research and biotechnology has dual effects. First, futuristic techniques allow malevolent actors to design, synthesize, and release engineered pandemics that could be far deadlier than COVID. Even without awful intention, insufficient biosafety standards may contribute to some accidental release of deadly pathogens, even if they were being researched for beneficial reasons. Second, it helped and will help the development of efficient vaccines and antiviral treatments to overcome the pandemic. As we all know, Katalin Karikó and Drew Weissman have been awarded the Nobel Prize in Physiology or Medicine 2023 "for their discoveries concerning nucleoside base modifications that enabled the development of effective mRNA vaccines against COVID-19" [7]. Which tendency, biotechnology's beneficial or malignant effects, will be more robust?

There are causal relationships among deforestation, wildlife trade, and the rise of infectious diseases [8]. Spillover of pathogens from animals to humans is identified as the major cause of emerging infectious diseases and as the primary cause of COVID-19. The legal and illegal trade of wildlife for pets, meat, or medicine increases transmission. It looks clear that wildlife trade should be much better controlled. National and international trade of high-risk species like primates, bats, pangolins, civets, and rodents should be banned. Although future pandemics may have higher mortalities, the probability of total human extinction is small.

(v) Artificial intelligence

New developments in the application of artificial intelligence (AI) research became known to the general public, mostly related to the extensive use of contemporary language models, but also in image, video, and speech recognition and autonomous robotics. The new results generated significant excitement. We college professors spend much time assessing whether a student's essay was generated purely by ChatGPT or as the result of meaningful individual effort. More seriously, the new AI transforms, among others, healthcare, finance, and manufacturing. The most potent method of contemporary AI research, a sophisticated version of multi-layer artificial neural networks, grew from one of the two Fathers of cybernetics, Warren McCulloch, as it was mentioned in Sect. 1.2.1.1 who constructed with his prodigy coworker Walter Pitts the framework of the first computational model of the brain and mind [11]. I am making a big jump to the present; my peers from the International Neural Network Society edited a volume (Artificial Intelligence in the Age of Neural Networks and Brain Computing) [12]. *Brain-like computing* provides the framework of *deep learning*, which is the basis of the present day's techniques mentioned in Sect. 1.2.1.1, see [14, 15].

Humans have a natural reason to worry about the misuse of new methods; nowadays, cybersecurity is a significant source of concern. The potential misuse of AI is a critical issue that requires careful analysis and regulation.

As AI systems become more powerful, it is easy to imagine they may become superior to human performance. Will AI help us to create prosperity, or will it destroy us? A recurring question is whether any forces may unintentionally or intentionally destroy human society. The year 2023 is the year of statements to defend our civilization.

(1) Leaders of the tech industry presented a warning statement:

> "Mitigating the risk of extinction from A.I. should be a global priority alongside other societal-scale risks, such as pandemics and nuclear war." [10]

(2) The Biden administration issued an "Executive Order on the Safe, Secure, and Trustwo rthy Development and Use of Artificial Intelligence" [13].

(3) The European Union issued an Artificial Intelligence Act that deals with comprehensive rules for trustworthy AI [16]. Some important items are:

- Limitations on the use of biometric identification systems by law enforcement.
- Bans on social scoring and AI used to manipulate or exploit user vulnerabilities.
- Rights of consumers to launch complaints and receive meaningful explanations.
- Fines range from 35 million euros or 7%

We do not yet see how the statements will be converted into actual actions, but cautious optimism suggests that democracies will figure out how to deal with AI.

To reduce the probability of existential risk due to AI, we should understand the *alignment problem* [17]. AI designers want to solve practical problems, and it is far from clear how they will (if at all) align with human goals. Well, it is impossible to define *human goals*: how would one precisely define desirable and non-desirable behaviors for all humankind? Can an AI designer define a goal reflected by an *objective function* to be optimized, which will serve humans, or will AI serve her interest?

As a member of the neural network community, I have been strongly influenced by Geoffrey Hinton's warning signal in an interview with the New York Times [18] about the existential risk AI may generate if misaligned. It is interesting to know that Norbert Wiener (one of the Fathers of Cybernetics, as the Reader may remember) already pointed out the problem: "If we use, to achieve our purposes, a mechanical agency with whose operation we cannot interfere effectively...we had better be quite sure that the purpose put into the machine is the purpose which we really desire." [19].

In a famous article, John von Neumann (1903–1957) asked the question *Can We Survive Technology?* [20]. He listed the dangers from nuclear energy via climate problems to automation. His words are still valid today:

> There is no cure for progress. Any attempt to find automatically safe channels for the present explosive variety of progress must lead to frustration. The only safety possible is relative and lies in an intelligent exercise of day-to-day judgment.

6.2   Complex Systems Approach to Political Instability

6.2.1   Predicting and Controlling Terrorism: Possibilities and Limitations

A subset of terrorists wants to save the world, while others want to destroy it. More precisely, revolutionary terrorists want to save the world by destroying the actual political system to substitute it with a different one. Red Army Faction [21] was a far-left terrorist group from the early nineteen seventies. They identified the older generations of Germans with Nazism, and they wanted to transform the orde. Now we have Hamas ... BBC has a reputation for being as objective as possible [22]:

> What is Hamas, and what does it want?
> Hamas is a Palestinian group that has run Gaza since 2007.
> The name is an acronym for Harakat al-Muqawama al-Islamiya, which means Islamic Resistance Movement.
> The group wants to destroy Israel and replace it with an Islamic state.

Between these two events, 9/11 occurred in 2001, when we learned again that events with small probabilities and massive impacts on society might happen, and that social systems are complex and often beyond the limits of predictability. These days, in addition to "conventional" terrorism, we should know about the options of bioterrorism, nuclear terrorism, and cyberterrorism.

While terrorists kill innocent people, at the systemic level, the single major issue is that terrorism undermines democratic institutions. It generates fear, which is terrible since it strengthens political extremes and further polarizes society.

There is a positive feedback mechanism to recruit terrorists. ISIS has been gaining its reputation through synchronized and well-advertised atrocities such as public beheadings. After its publicity increased, more and more people were recruited through social media, predominantly young male Muslims.

There is a control mechanism to cope with terrorism. Computational social science may help recognize early signals, the first step in any negative feedback strategy to preserve stability. However, it is technically challenging since there is a huge number of *potential* terrorists, but the cardinality of the set of *actual* terrorists is small. *How to (better) find a perpetrator in a haystack*, asked the authors of an excellent paper [23]. At the same time, the application of new machine learning improved the efficacy of the screening procedure. According to this approach, primarily by reducing the haystack size rather than on the identification of the needle.

Modeling the risk of terrorism

Adopting the spirit of applying system dynamics to social systems discussed in Sect. 1.2.3, a causal loop of a skeleton model of terrorist dynamics can be given [24], see Fig. 6.1.

This figure shows the variables that drive changes in an insurgent population and the causal relationship among them, and the signs reflect the positive and negative

Fig. 6.1 Skeleton model for
the dynamics of a terrorist
population. Based on the
logic of [24]

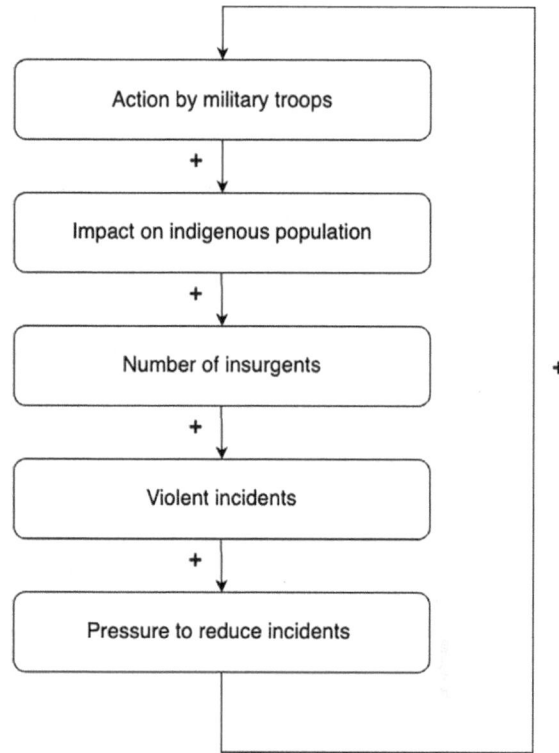

influences. The narrative of the diagram is as follows. Any change in the number
of insurgents causes a change in the number of violent incidents. The increase in
the number of insurgents forces the government to take action to decrease incidents.
These actions cause responses by military troops, which leads to a change in how the
"indigenous" population responds. Their response then contributes to the number
of people participating in the insurgency. The model reminds us of epidemiological
models, in which susceptible members are converted to be infected (and ineffective!).
A similar scenario occurs when passive people are recruited to be terrorists. Simu-
lating the velocity of terrorist recruitment requires estimating the numerical values
of state variables and constants, which regulate the velocities of the subprocesses.
Again, our best chance to get information about the risk level of terror is to make
more realistic models and simulate possible scenarios. One critical question is to
know the counter-terrorism measures that would be sufficient to defeat the terrorist
organization.

Terrorism alone is generally not considered an existential risk on a global scale,
as it typically does not pose a threat to the survival of humanity as a whole. However,
the impact of terrorism can vary. Terrorists not only kill people, they poison souls by
generating fear.

6.2.2   Is the Offense-Defense Balance Stable?

The offense-defense balance: the data

- The website *Our World in Data* [25] maintained by Max Roser from Oxford has a prominent graph showing that per-capita deaths from war do *not* show a long-term trend over the last six hundred years. The same result occurred for wars with horses and swords than tanks and war-planes. Historical data suggest that the phenomenon called the *offense-defense balance* exists, so there should be some stabilizing mechanism.

The offense-defense balance was discussed recently [26], posed several questions, and offered some (not necessarily final) explanations:

(i) Will future AI change the balance for helping support better either offense or defense? One possible explanation is that the expected significant future changes in AI at the system level are somewhat analogous to the rapid transitions during the Industrial Revolution. One difference is that, in principle, a small group of malignant hackers and terrorists endowed with appropriate technology could destroy the internet and maybe countries and whole continents. So we have the following question:

(ii) Did the pattern of the offense-defense balance change in the age of cyber-wars and cybersecurity? Since there are newer and newer attack techniques, it is not enough for defenders to detect and respond to offense; they also should use techniques of prediction and prevention. Traditional defense strategies based on a rigid list of $IF \rightarrow THEN$ actions should be changed by learning adaptive strategies to understand hackers' mindset [27]. Hackers use as much automation as possible to boost their attacks. Therefore, defenders should find strategies for repeated large-scale attacks. Hackers try to attack less fault-tolerant points, so defenders should know their weakest points well.

(iii) What can we expect in the age of developed biotechnology? Can the biological attack be transferred from action movies to reality? (I am not very familiar with this genre, so here is a list of *10 Best Biological Warfare Movies To Watch, Ranked According To IMDb* prepared several years ago [28]). Offenders may adopt bio-weapons by intentionally releasing toxic biological substances that cause various diseases. Disease substances, such as viruses, can be transmitted from person to person and may lead to high mortality and imply severe disruption of society. To estimate the probability of threats, we should know the different scenarios. Biological threats may be naturally occurring, accidental (e.g., released unintentionally from a lab), or deliberate. Probably and hopefully, COVID-19 belongs to the first category and did not escape from a scientific research lab.

As we learned well during the COVID era, vaccines are intended to prepare our immune system to fight against infections and prevent disease. Protective vaccines are critical components of strategies to cope with the impact of bioweapons threats. International cooperation is an absolute must among research institutions, pharmaceutical companies, and administrative organizations to accelerate the development of protective vaccines [29]. The U.S. Department of Defense issued a review of the

Biodefense reform [30] and determined the goals of building early detection and warning systems to give rapid answers to limit the impacts of big incidents.

(iv) Towards some explanation: Offenders are also defenders and vice versa

It is impossible to estimate the likelihood that a lunatic madman or madwoman will produce some gigantic catastrophe. While the emergence of advanced AI may imply colossal changes, feedback loops may stabilize the balance. There is no reason to assume that AI will prioritize the acceleration of the development of offenders or defenders.

Bioweapons versus vaccines in the age of AI

Biological warfare has a history [31] and was used in Asia during World War II. Bacteria that cause anthrax are one of the most likely candidates to be used in a biological attack.

What can we expect in the age of new AI? OpenAI's Sam Altman has called for regulation on AI models "that could help create novel biological agents" [32]. Databases of viruses can be expanded greatly by using AI techniques, which could generate new chemical knowledge of new viruses. New AI models could be trained to find new research pathways in areas like pathogens and cancers; we cannot see how to control them to design newer and more harmful pathogens. DNA synthesis companies should be "screened as in compliance with U.S. regulations prohibiting the possession, use, and transfer of specific pathogens and biological toxins" [33].

So, should we be optimistic or pessimistic as we face the development of new technologies? Recent debates exist among *techno-optimists* and *techno-pessimists*, as two Manifestos reflect [34, 35].

Techno-optimism, techno-pessimism and the existential risk

Techno-optimism is the view that technological development improves human life through new possibilities and solutions to existing or future problems. Conversely, techno-pessimism does not believe that technological development implies real progress in the quality of human life and worries about its unintended and uncontrollable consequences. Techno-pessimists blame the Industrial Revolution as the primary cause of climate change due to the emergence of fossil fuels and predict that more destruction may happen by transitioning to a renewable energy system.

The two camps offer different approaches to managing global problems, such as climate change. Techno-optimists emphasize technology-based solutions, while techno-pessimists suggest changes in people's and society's behavior, such as carbon tax and degrowth.

"Degrowth is an idea that critiques the global capitalist system, which pursues growth at all costs, causing human exploitation and environmental destruction. The degrowth movement of activists and researchers advocates for societies prioritizing social and ecological well-being instead of corporate profits, over-production, and excess consumption. This requires radical redistribution, a reduction in the material size of the global economy, and a shift in common values towards care, solidarity, and autonomy. Degrowth means transforming societies to ensure environmental justice and a good life for all within planetary boundaries." [36]

It is essential to find a balanced view between the extreme perspectives. It would be difficult to deny that technology has improved the human condition, health, prosperity, and quality of life worldwide. Technological progress is undoubtedly not the omnipotent medicine for all human problems, but it looks better than other alternatives.

However, techno-optimists should not underestimate the dangers of development. Our best hope is that feedback control techniques find the trajectories towards prosperity and will avoid risk. On the one hand, the optimism is not baseless: reason, science, and humanism are the most vital features of humanity. On the other hand, the concept of feedback control is used in a somewhat vague metaphorical sense. We do not have to think giving some supermodel and using some optimal control theory is possible. However, knowing the causal loops among subsystems gives a tool to avoid at least huge catastrophes. I might be wrong.

6.2.3 How Much Social Unrest Do We Need?

Social or civil unrest, also known as civil disorder, can be identified with situations when the public order is disturbed, and the active involvement of law enforcement is necessary to keep/restore the "normal" life. Social unrest occurs when a group gathers publicly to express anger or dissatisfaction focused on a common issue. Social unrest is frequently a feedback process: when people perceive systemic inequality, discrimination, or injustice, it can mobilize *collective action* to respond to these issues. Collective behavior is the emotional reaction of the crowd members due to fears and prejudices. It is driven by social factors such as the strengths or weaknesses of leadership and moral issues. Psychological factors are also involved, e.g., people may imitate each other. Emotional contagion may also occur, which involves the spontaneous spread of emotions, which can happen by a positive feedback mechanism from one person to another or in a larger group.

Looking back to the not-too-distant past, protests against the Vietnam War started in the mid-sixties. The late sixties were a period of high racial tension in the US. During the summer of 1967, disorder broke out in many cities. The bloodiest of the urban riots was in Detroit in the summer of 1967. Twenty-five years later, thousands of people participated in the Los Angeles riot over a week. Everybody remembers a series of police brutality protests that began in Minneapolis on May 26, 2020. The civil unrest and protests evoked international reactions to the murder of George Floyd, a 46-year-old African American man, during an arrest [38].

According to the Global Protest Tracker data of the Carnegie Endowment for International Peace [39], 135 significant economic antigovernment protests have occurred since 2017. (Please note: numbers are dangerous; they have both algebraic meaning and emotional charge.) More often than not, demonstrators have a collective "now or never" feeling that forces them to act.

How much social unrest do we need? Zero social rest would reflect a very static society. Too much unrest would make the continuity of the normality impossible. But sometimes people realize that "normality," whatever it means, can not be maintained, and social unrest may turn into revolution. Sudden changes in political, economic, and cultural institutions characterize a successful revolution. It may remind the Readers of the general phenomenon called bistability mentioned in Sect. 3.3.2, a sudden transition from one state to another.

The *Armed Conflict Location and Event Data Project* [37] provides a database, maps, and analysis about actors, dates, fatalities, locations, and locations of reported political violence and demonstration events worldwide. All types of political agents, such as governments, rebels, militias, identity groups, political parties, external actors, rioters, protesters, and civilians, are considered. It is very detailed and contains 1.3 million individual events globally.

The source of civil disorder is often political grievances ranging from a simple protest to a mass civil disobedience. Sometimes, these events can be spontaneous, but they can also be planned. Mutual excitation between agitators and law enforcers forms a positive feedback loop and might lead to uncontrolled violence. An example is "news from today" (January 9th, 2024):

> "Ecuador descended into chaos this week as an influential gang leader disappeared from jail, uprisings broke out in several prisons, and guards were kidnapped and threatened by inmates in what has quickly escalated into a major crisis for the South American country.
>
> The unrest continued on Tuesday afternoon when masked men stormed a television station in Guayaquil, the nation's largest city, taking anchors and staff hostage and exchanging gunfire with the police as cameras rolled. The standoff ended after the police subdued and arrested the intruders." [40]

Civil disorder has arisen from different economic and political reasons. The *Occupy movement* protests emerged from the perceived social and economic inequality and the lack of real democracy around the world. Its main slogan was "We are the 99%". While social unrest implies disrupting public order, which may threaten public safety and reduce GDP, it might also be a driving force toward beneficial reforms. Social epidemic models describe the propagation of ideas and the emergence of collective actions. The endogenous positive feedback process is a critical driving force behind social riots. Too much stability conserves the social status quo; too much instability is unbearable.

Predicting social unrest

A significant progress in interdisciplinary science is the interaction between social science and policy-making with data science and, more generally, computational science. Specifically, recent development in collecting data and adopting machine learning techniques helps to make forecasts of social unrest one year ahead [41].

(i) Data collection

A Reported Social Unrest Index (RSUI) was constructed using media reports [42]. The source is a massive set of English-language newspapers, and the index is given by the number of articles with the phrases *protest* or *riot* or *revolution* or *unrest*. Unrest is very different and can be classified as Government, Democratic-reforms, Elections, Global issues, Religious, Basic Needs, Coups, and Violence. Items in the individual categories may overlap.

The Government category contains protests relating to presidents, political opposition, and political coalitions. Democratic reforms are broader, including protests around democratic reforms or rights, corruption scandals, and the free press. Global issues motivate people across many countries, such as climate and, more generally, environmental issues, but they also include anti-war, anti-globalization, and anti-immigration unrest.

(ii) Predictors

What types of data should be used as indicators of making predictions?

First is poverty, including food insecurity. Second is access to essential services such as electricity, education, and health care. Third is financial and social inequality, including gender inequality. Fourth is population density since the propagation of ideas happens more effectively in densely populated urban environments than in rural areas. Fifth, unemployment, mainly youth unemployment, is also a significant indicator. The family of indicators is much more detailed; here, only the essential components were listed.

(iii) Modeling

Different machine learning methods were used, including, but not exclusively, neural network techniques. To validate a prediction method, some backtesting should be done. Here, an initial training set of annual data from 1990 to 1995 was used to predict the test set of 1996. Then, it rolled forward, generating predictions up to the year 2019.

(iv) Insights

The model produced a probability of unrest for each country for each year. This probability can be identified as a risk index—an indicator of the likelihood of unrest in the following year. The accuracy level of the models is 66%, so it is significantly better than pure chance (50%). Social unrest implies financial, economic, and political risks in the short term and might have potentially positive effects over the medium and longer term. Permanent large-scale unrest generates too much loss and uncertainty, while the lack of unrest means no social feedback for the governing institutions and people. Obviously, *social engineering* [43] is not an option. Social engineers from the late 19th and early 20th century placed their expertise and knowledge above mass democratic decision-making. But the spirit of adaptive control [44], at least in a metaphorical sense, might be helpful. A controller uses a control method that must adapt to a controlled system with varying parameters in uncertain situations. Thinking about the nature of this control may help participants and decision-makers to manage the degree of social unrest.

6.2.4 How Much Migration Do We Need?

Migration, the movement of people from location A to location B, is not unnatural. It can be voluntary or involuntary. Location A might be unfavorable for various reasons, including economic, environmental, political, and social issues, and repel people to leave this place. Location B might be more attractive for the same reasons and offer the hope of a better life. As we learned from the Moana movie mentioned in Sect. 2.1.2, the Polynesians had to leave their island as living conditions became sour. By adopting early control technology, they managed to migrate to better places.

Migration can have both stabilizing and destabilizing effects.

Stabilizing effects

(i) *Demography.* Countries with higher emigration rates (of the more mobile younger generation) have an aging population with a lower birth rate. Immigration may stabilize the more appropriate demographic structure necessary for a well-operating workforce.

(ii) *Economics.* As a direct consequence, the migrants not only accept jobs that local people do not take, but might also increase innovation and productivity by exporting their specific knowledge.

(iii) *Diversity.* People from diverse cultures may contribute to new ways of thinking, new knowledge, and different experiences. Initial negative effects were driven by a reduction in the trust of others around them in countries with increased diversity. However, after a period, the positive effect of mixing with members of different groups may overcome the initial negative effects [32].

Destabilizing effects

Disturbance in service At least in the short term, rapid mass migration may imply a shortage in public service (housing, education, health service, etc.).

Initial social tension. It is generally the first warning signal that communities feel threatened by the loss of jobs and cultural homogeneity.

Economics. Migration of a more educated workforce from developing countries to more developed ones amplifies the already existing economic in-homogeneity since the human capital necessary for development is reduced.

Political instability Migration issues proved politically divisive, leading to policy challenges and potential instability. Opinions over immigration policies may further increase the polarization of societies.

Keeping the balance between maintaining control over migration and respecting human rights and international obligations needs a balanced approach. In one of the most influential books about immigration, [51] a Cuban-American, Harvard labor economist George J. Borjas argues "I am an immigrant, and yet I do not buy into the notion that immigration is universally beneficial …But I still feel that it is a good thing to give some of the poor and huddled masses, people who face so many hardships, a chance to experience the incredible opportunities that our exceptional country has to offer."

Causal modeling combined with data processing methods may help predict migration and contribute to providing an effective migration control policy.

Predicting migration

(i) Data collection

The European Agency for Asylum collects data about asylum-related migration and informs the public about recent trends [52]. In 2023, the EU received more than one million asylum applications, close to the same as during the 2015-16 refugee crisis. The United Nations Global also established and maintains a migration database. It collects data on the number of international migrants by country of birth, citizenship, and sex from more than 200 countries and territories worldwide. Both stock and flow data are stored (interestingly in the spirit of system dynamics seen in Sect. 1.2.3).

(ii) Predictors

Both traditional causal models and newer machine-learning techniques are used to predict international migration. Prediction is a necessary step before thinking about control strategies. Local conflicts, epidemics, environmental catastrophes, and poverty have been identified as the main drivers of forced migration [53].

Migration can be conceptualized as a two-step process. First, some factors push people to leave their homes. Second, more careful decision-making involves a more systematic search process for potential destinations based on some cost-benefit analysis. At the same time, attractiveness is typically estimated purely by the per capita income difference between the destination and origin, the culture of the receiving country. Historically, Canada is viewed as welcoming of immigrants, as seen by an index constructed by Gallup (Migrant Acceptance Index) to assess people's acceptance of migrants, followed by Iceland and New Zealand [54].

(iii) Modeling

The first traditional model framework of human mobility, the so-called *gravity model* (for a relatively early application, see [55]) is based on the assumption that the probability of a transition between locations A and B decays as a function of the distance between them, and all the other factors are neglected.

More recently, different versions of a *radiation model* have been suggested [56] to capture long-range trips better than gravity-based models. The model assumes that the probability of transition directly depends on more general opportunities and less directly on physical distance. There is always a trade-off between assumptions-based direct models and data-driven machine learning methods. While direct models better serve the *understanding* of the causal relationships, with the available data, machine learning methods seem more flexible and can parameterize for different countries or regions.

(iv) Insights

Data do not necessarily support the argument that we live in a world of uncontrollable migration [57]:

> The vast majority of people who migrate do so voluntarily and without drama. For all the talk of record numbers and unprecedented crisis, the share of the world's people who live outside their country of birth is just 3.6%; it has barely changed since 1960 when it was 3.1%. The numbers forcibly displaced fluctuate wildly, depending on how many wars are raging, but show no clear long-term upward trend. The total has risen alarmingly in the past decade or so, from 0.6% in 2012 to 1.4% in 2022. But this is only a sixth of what it was in the aftermath of the Second World War.

The outcome of migration influences from A to B is a feedback signal for the future migration flow. It is a flow of information sent to the area of origin about the success or failure of specific migration projects. Feedback may amplify or reduce the flow: further migration may be encouraged or extinguished from movement [58].

A society that implements some trade-off between too homogeneous and too diverse looks like the best arrangement. Advantageously, homogeneous societies tend to have a common set of cultural and moral rules. People have a common sense of what to expect in most situations. However, they tend to be less creative because they all have the same background, experience, and knowledge.

6.2.5 Why Do Democracies Do Better than Autocracies? A Systems-Theoretical Analysis

A systems-theoretical framework of political systems was suggested by David Easton (1917–2014) [60], where the relationship between the system and its environment was emphasized, see Fig. 6.2.

More specifically, there are the policymakers who make decisions and the public who reacts to these decisions. Policymakers are problem-solvers. Due to their limited capacity of cognitive resources, decision-making contains errors that are impossible to avoid. How do politicians deal with these errors? There are two extreme cases. A positive feedback mechanism leads to *error accumulation*, which may generate crises, such as increased gun violence rate, banking crisis, famine, etc., and *error-correcting* mechanisms implemented by negative feedback. The Punctuated Equilibrium Theory [61] explains both slow, gradual, and sudden large changes.

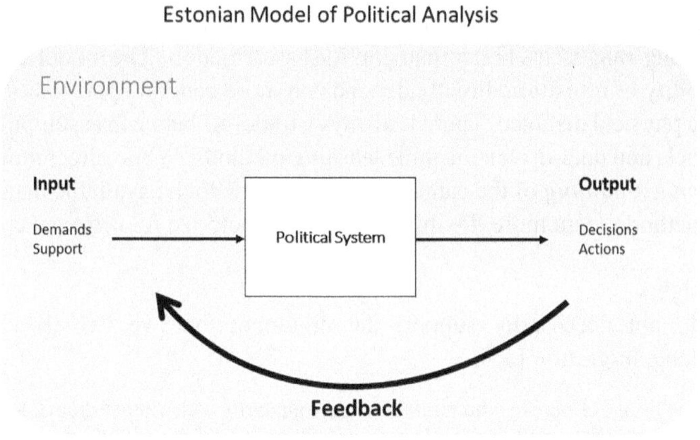

Estonian Model of Political Analysis

Fig. 6.2 The Estonian model of policy analysis

Since the thermostat is the textbook example device for implementing negative feedback, it is unsurprising that a *thermostatic model* of public preferences has been suggested [62, 63]. The model's primary assumption is that the public sends policymakers feedback signals. For any agenda (say, military expenses), the signal may be preferences for "more" or "less" spending.

The model suggests that in democratic institutions, there is a bidirectional information flow between the public and policymakers. The public acts as a thermostat and informs policymakers about their preferences for increasing or decreasing the temperature. Policymakers serve as a feedback unit, responding to the requested changes. Like every well-functioning model, it has scope and limits but is simple and illuminating.

Policy disasters are defined as "avoidable, unintended extreme negative policy outcomes" [64]. The working hypothesis was that systems with higher error accumulation produce more policy disasters. Now, there are data available to test the hypothesis. Data on financial crises and natural and technological disasters across 70 countries over 60 years were analyzed. Systems with weaker information flow and more veto players tend to have greater policy disaster risk. (A veto player is a political actor who can unilaterally decline a choice.)

Do democracies perform better than more autocratic political systems? Ten years ago or so, the predominant view was that closed autocracies, where veto players and information flows are constrained, have a risk of policy disasters. Liberal democracies have the lowest risk of policy disasters.

I corresponded with Bryan Jones, a leading political scientist at the University of Texas at Austin (I am proud of being his coauthor in a paper). As he writes:

> Today, there is an upswing in support for oligarchies, and the autocrats claim they are good at solving problems. They are not mainly because they are terrible at processing information and defining problems. They accumulate errors. Democracies are better, and some structures of democratic rule are better than others in doing so. The US system is not the best because

there are too many *veto points*, as we political scientists term them. The continual adjustment required to implement error adjustment (negative feedback) is limited because of the structure of the government.

Adaptation is a crucial mechanism to deal with changes. However, fully rigid policies in a highly controlled political system may provide unchanging policies over long periods. Still, they are generally not responsive to the changing nature of the policy environment. In other words, they are less adaptive systems. We should mention the time scales here: One reason today's situation might be much worse is that change is happening much more rapidly than ever before, and most institutions, particularly governments, cannot adjust fast enough.

A famous example is the Chinese famine associated with Mao's Great Leap Forward [65]. It was the worst famine in modern history, causing approximately 30 million starvation deaths. Policies based on collectivization, industrialization, and resistance to information implied great suffering in the countryside. The information flow among the different levels of the Communist Party hierarchy was weak or false. They frequently reported exaggerated harvest estimates to impress superiors. The government did not have any error-correcting mechanism. History shows that the famine ended with policy changes that began in 1961, as the government implemented reforms. But this only happened after extensive and needless suffering.

6.3 Lessons Learned and Looking Forward

In this chapter, in the beginning, we discussed the scopes and limits of the possibilities to estimate the probabilities of existential risk due to natural and man-made reasons. In his book, the system scientist and excellent science writer John Casti offers several scenarios for the "collapse of everything" [45]. He explains that contemporary interconnected complex societies have become highly vulnerable to extreme internal or external events that will ultimately destroy civilization. While I adore Casti's style and knowledge, instead of seeing a catastrophic future, I try to adopt a well-informed but cautiously optimistic perspective. The process is very complex, and it is challenging to offer feedback control strategies. It is difficult but not impossible. The concept of "feedback control strategies" here should be understood metaphorically. We are far from applying formal models (e.g., [47]).

Handling Social Complexity. A methodology has been suggested [46] based on three major components for handling complex problems: knowledge, power, and emotion. Knowledge is needed to build a project, and this is high-quality technological knowledge for large technological engineering projects. Power focuses on how actors can stimulate, delay, or prevent a project. Emotion refers to the psychological aspects of human problem handling, i.e., intuition, likes and dislikes, trust, and fear. Handling complex problems involves many elements that cannot be anticipated beforehand. Therefore, a methodology for guiding complex problems offers only frameworks, maybe simple skeleton models.

The role of feedback is evident in any complex system. Positive feedback mechanisms generate a "rich-get-richer" phenomenon, often leading to skewed (power law) distributions. (See, e.g., [48]) for a brief history of producing such distributions.) Among others, an article *What science can do for democracy: a complexity science approach* [50] tries to answer the ambitious question. As far as social systems are concerned, economic inequality can readily be converted into unequal political power. The implication of increased wealth inequality is the role of (very) wealthy actors to change or sustain the color of the political institutions. Wealth can be translated into increased political power. If you write a book, it is better to avoid mentioning actual political circumstances. In any case, there is a self-reinforcing feedback loop between a strong presidential candidate, his base, supportive media, and politicians who continue to support his line [49]. A super wealthy man may owe an influential social media and can propagate lies say, about the voting rights of immigrants. He and other players shift the struggle between authoritarianism and democracy towards transforming the society into an effective oligarchy. We see the loop. High inequality will motivate unhappy citizens to mobilize themselves, leading to anti-inequality demonstrations and social unrest. In some European countries, elected governments pressured the media to occupy key, in principle, independent institutions such as courts. (The Reader can guess which countries are in the author's mind). The stability of democratic institutions seems to be in danger.

According to an overused Churchill's quote: "Indeed it has been said that democracy is the worst form of Government except for all those other forms that have been tried from time to time." When I decided to choose the maybe too bombastic subtitle ("How to Destroy or Save the World"), a critical item in my mind was how to avoid the progress of democratic backsliding, and it is a process when autocratic leadership reduces the role of democratic institutions. We need a stabilizing feedback reaction. It is a normative statement (mentioned in Sect. 5.6). My message is that we may hope to learn from feedback control theory how to implement working stabilizing mechanisms.

References

1. Servigne P, Stevens R (2020) How everything can collapse. Cambridge, Polity. ISBN 9781509541393
2. Diamond J (2005) Collapse: how societies choose to fail or succeed. Penguin Books. ISBN 978-0-241-95868-1
3. Cotton-Barratt O, et al (2016) Global catastrophic risks 2016, Annual Report, Global Priorities Project. https://globalchallenges.org/app/uploads/2023/06/Global-Catastrophic-Risks-2016.pdf
4. Robock A (2010) Nuclear winter. Clim Change 1:418–427
5. https://www.statista.com/statistics/264435/number-of-nuclear-warheads-worldwide/
6. https://www.ceres.org/homepage
7. https://www.nobelprize.org/all-nobel-prizes-2023/
8. Dobson AP, Pimm SL, Hannah L, Kaufman L, Ahumada JA, Ando AW, Bernstein A, Busch J, Daszak P, Engelmann J, Kinnaird MF, Li BV, Loch-Temzelides T, Lovejoy T, Nowak K,

Roehrdanz PR, Vale MM (2020) Ecology and economics for pandemic prevention. Science. https://doi.org/abc3189

9. https://www.quora.com/What-if-we-could-control-natural-disasters
10. Statement on AI Risk. https://www.safe.ai/statement-on-ai-risk
11. McCulloch WS, Pitts W (1943) A logical calculus of the ideas immanent in nervous system. Bull Math Biophys 5:115–133
12. Kozma R, Alippi C, Choe Y, Morabito FC (2023) Artificial intelligence in the age of neural networks and brain computing. 2nd edn
13. Executive order on the safe, secure, and trustworthy development and use of artificial intelligence. The White House, Oct 30, 2023 https://www.whitehouse.gov/briefing-room/presidential-actions/2023/10/30/executive-order-on-the-safe-secure-and-trustworthy-development-and-use-of-artificial-intelligence/
14. LeCun Y, Bengio Y, Hinton G (2015) Deep learning. Nature 521:436–444. https://doi.org/10.1038/nature14539
15. Schmidhuber J (2015) Deep learning in neural networks: an overview. Neural Netw 61:85–117. arXiv:1404.7828. https://doi.org/10.1016/j.neunet.2014.09.003
16. https://www.europarl.europa.eu/news/en/press-room/20231206IPR15699/artificial-intelligence-act-deal-on-comprehensive-rules-for-trustworthy-ai
17. Christian B (2020) The alignment problem: machine learning and human values. WW Norton
18. Metz C. The Godfather of A.I. Leaves Google and Warns of Danger Ahead, https://www.nytimes.com/2023/05/01/technology/ai-google-chatbot-engineer-quits-hinton.html
19. Wiener N (1960) Moral and technical consequences of automation, science, new series, vol. 131, No 3410, pp 1355–1358
20. Von Neumann J (1955) Can we survive technology? +++
21. https://en.wikipedia.org/wiki/Red_Army_Faction
22. What is Hamas and why is it fighting with Israel in Gaza? https://www.bbc.com/news/world-middle-east-670399
23. Neuman Y, Cohen Y, Neuman Y (2019) How to (better) find a perpetrator in a haystack. J Big Data 6, 9. https://doi.org/10.1186/s40537-019-0172-9
24. Ezell BC, et al (2010) Probabilistic risk analysis and terrorism risk. Risk Anal. 30(4). https://www.dhs.gov/xlibrary/assets/rma-risk-assessment-technical-publication.pdf
25. https://ourworldindata.org/
26. https://forum.effectivealtruism.org/posts/oHkJziKoP47PBSizr/the-offense-defense-balance-rarely-changes
27. Why defenders should embrace a hacker mindset. The Hacker News. Nov 20, 2023. https://thehackernews.com/2023/11/why-defenders-should-embrace-a-hacker.html
28. Natasha_B: 10 best biological warfare movies to watch, ranked according to IMDb. Screenrant https://screenrant.com/best-biological-warefare-movies-ranked-imdb/#the-satan-bug---6-2
29. Pedro J (2023) Protective vaccines against diseases linked to potential bioweapon threats. J Vaccine Vaccination. https://www.walshmedicalmedia.com/open-access/protective-vaccines-against-diseases-linked-to-potential-bioweapon-threats.pdf
30. 2023 BIODEFENSE POSTURE REVIEW (BPR) https://media.defense.gov/2023/Aug/17/2003282337/-1/-1/1/2023_BIODEFENSE_POSTURE_REVIEW.PDF
31. Frischknecht F (2003) The history of biological warfare. EMBO Rep 4(Suppl 1):S47–52, https://doi.org/10.1038/sj.embor.embor849
32. Hendrix J (2023) Transcript: senate judiciary subcommittee hearing on oversight of AI. Tech Policy Press, https://techpolicy.press/transcript-senate-judiciary-subcommittee-hearing-on-oversight-of-ai/
33. https://www.federalregister.gov/documents/2020/08/26/2020-18444/review-and-revision-of-the-screening-framework-guidance-for-providers-of-synthetic-double-stranded
34. Andreessen M (2023) The techno-optimist manifesto. https://a16z.com/the-techno-optimist-manifesto/
35. Yarvin C (2023) A techno-pessimist manifesto. https://graymirror.substack.com/p/a-techno-pessimist-manifesto

36. https://degrowth.info/degrowth
37. Armed conflict location and event data project, https://en.wikipedia.org/wiki/Armed_Conflict_Location_and_Event_Data_Project
38. How George Floyd was killed in police custody, https://www.nytimes.com/2020/05/31/us/george-floyd-investigation.html
39. https://carnegieendowment.org/publications/interactive/protest-tracker
40. Ecuador Plunges Into Crisis Amid Prison Riots and Gang leader's disappearance. https://www.nytimes.com/2024/01/09/world/americas/ecuador-gang-prison-emergency.html
41. Redl C, Hlatshwayo S. Forecasting social unrest: a machine learning approach. Volume/Issue: Volume 2021: Issue 263 Publisher: International Monetary Fund, https://www.elibrary.imf.org/view/journals/001/2021/263/article-A001-en.xml
42. Barrett P, Appendino M, Nguyen K, de Leon Miranda J (2020) Measuring social unrest using media reports. IMF Working Paper No. 20/129
43. Gehl RW, Lawson ST (2022) Social engineering: how crowdmasters, phreaks, hackers, and trolls created a new form of manipulative communication. The MIT Press
44. Egardt B (1979) Stability of adaptive controllers. Springer, New York
45. Casti J (2012) X-events: the collapse of everything. William Morrow
46. DeTombe D (2015) Handling societal complexity. A study of the theory and the methodology of societal complexity and the COMPRAM methodology. Springer, Heidelberg
47. Åström KJ, Murray RM (2008) Feedback systems: an introduction for scientists and engineers. Princeton University Press
48. Mitzenmacher M (2004) A brief history of generative models for power law and lognormal distributions. Internet Math 1(2):226–251
49. Trump – and the rule of law – on trial. International Bar Association. https://www.ibanet.org/Trump-and-the-rule-of-law-on-trial
50. Wiesner K, et al (2019) Stability of democracies: a complex systems perspective. Eur J Phys 40:014002
51. Borja GJ (2016) We wanted workers: unraveling the immigration narrative. W. W. Norton & Company, Illustrated edition
52. https://euaa.europa.eu/latest-asylum-trends-asylum
53. Qi H, Bircan T (2023) Modelling and predicting forced migration. PLoS One 18(4):e0284416. https://doi.org/10.1371/journal.pone.0284416. PMID: 37053198; PMCID: PMC10101513
54. https://www.statista.com/chart/10804/the-countries-most-and-least-accepting-of-migrants/
55. Clark GL, Ballard KP (1980) Modeling out-migration from depressed regions: the significance of origin and destination characteristics. Environ Plann A 12(7):799–812
56. Simini F, González MC, Maritan A, Barabási AL (2012) A universal model for mobility and migration patterns. Nature 484, 7392:96–100
57. Leaders: how to detoxify the politics of migration. https://www.economist.com/leaders/2023/12/20/how-to-detoxify-the-politics-of-migration
58. Bakewell O, Kubal A, Pereira S (2016) Introduction: feedback in migration processes. In: Beyond networks. Feedback in international migration. Palgrave Macmillan London
59. Jones BD, Epp DA, Baumgartner FR (2019) Democracy, authoritarianism, and policy punctuations. Int Rev Public Policy, https://api.semanticscholar.org/CorpusID:195731191
60. Easton D (1953) The political system, an inquiry into the state of political science. Knopf
61. Baumgartner FR, Jones BD (1993/2009) Agendas and instability in American politics (1st, 2nd ed.). The University of Chicago Press
62. Wlezien C (1995) The public as thermostat: dynamics of preferences for spending. Am J Polit Sci 39(4):981–1000
63. Soroka SN, Wlezien C (2010) Degrees of democracy: politics, public opinion, and policy. Cambridge University Press
64. Fagan EJ (2021) Political institutions, punctuated equilibrium theory, and policy disasters. Policy Stud J 2023;51:243–263. https://doi.org/10.1111/psj.12460
65. Branigan T (2013) China's great famine: the true story. The Guardian

Chapter 7
Epilogue: The Narrow Border Between Prosperity and Destruction

Abstract Do we live in the best possible world or the shadow of existential risk? The answer is BOTH! To avoid disasters, we need to apply the spirit of feedback control.

Do we live in the best possible world or the shadow of existential risk?

The answer is BOTH! Steven Pinker, the rock star of cognitive science, states that life, wealth, well-being, security, education, knowledge, and happiness are improving in the West and worldwide [1]. Enlightenment values of reason, science, and humanism have brought progress and prosperity. The Oxford philosopher Toby Ord states that the extinction of the human race is the worst thing that could happen. In fact, until the last few years, the scientific and political society has devoted little study to the possible causes of extinction and the best ways to address them [2].

Extinction panic is not new but also not continuous. It is interesting to see [3] that about a hundred years ago, the public felt somewhat similar apocalyptic fear. The 1920s was after a big war and pandemic, a time of big unpredictable technological development. I agree that studying the history of extinction panic offers us strategies to face the problems of our age. In the last several years, we have seen a very productive research period providing significant results in understating the possible mechanisms of the emergence of natural and man-made risks, assessing the probabilities of their occurrence, and risk management strategies for reducing these probabilities.

The spirit of complex systems theory, specifically related to abrupt changes from one regime into another around critical states [4], implicitly suggests that there is a narrow borderline between the regimes of prosperity and destruction: "Although the complexity of systems such as societies and ecological networks prohibits accurate mechanistic modeling, certain features turn out to be generic markers of the fragility that may typically precede a large class of abrupt changes." Fine-tuned control may help to avoid real crises. But here is the critical question:

How could we avoid catastrophic scenarios and keep the world on a sustainable trajectory?

© The Author(s), under exclusive license to Springer Nature Switzerland AG 2024   111
P. Érdi, *Feedback*,
https://doi.org/10.1007/978-3-031-62439-1_7

The term *point of no return* emerged in air navigation and refers to the time and/or location during a flight when the aircraft no longer has enough fuel to return to its departure location. It is often used metaphorically to characterize irreversible changes at many levels, from individual life stories via technological evolution to climate changes. Malcolm Gladwell championed the expression *tipping point* [5]. The innovation trajectory may become unstoppable regarding technology, propelling humanity into uncharted territory.

Feedback control is not a silver bullet; however, its role everywhere is remarkable. We should remember its basic techniques even though humanity faces global problems beyond the limits of formal methods. It would be difficult to deny that we should be able to discriminate between processes we can and cannot control. We all now know that humanity has a to-do list to avoid catastrophes, and here is a very tentative list :

- Mitigate climate change
- Maintain the environment to support health and well-being
- Balance overproduction and degrowth
- Improve food redistribution methods and balance food waste versus insecurity
- Ensure social stability while also providing conditions for social development and progress

How should we characterize a resilient society?

It is not a new idea, but it is still good to remember that closed societies built on top-down, hierarchical systems of control and rule-based order try to maintain the status quo [6]. Generally, an ideological dogma or belief system provides the underlying operational framework. The goal is to act based on planning. Open societies accept the co-occurrence of both order and uncertainty. They try to unify constancy and change. New ideas naturally create uncertainties. It is good to assume that there is no perfect order and no absolute truth, not even as an asymptotic goal. A new problem is seen as an adverse event in a closed society. In an open society, it is seen as an opportunity for renewal. Somewhat paradoxically, open societies are, therefore, more stable. Autocratic closed systems show more instability (e.g., higher frequency of riots).

Cybernetics secretly returns

Motivated by the spirit of Wiener's dream, cybernetic concepts have been applied to understanding and modeling social systems. Financial markets, social and political dynamics, and online communities are the subject of analysis. While detailed mathematical models help social scientists and policymakers understand data, make predictions, and bring science-based decisions, we should accept that we do not have any supermodels. To avoid disasters, we need at least a three-component system: (i) early warning systems, (ii) decision-makers, and (iii) actuators.

Early warning systems may signal that the undesired amplification of initially small deviations has started and could lead to disasters. Rapid and local interventions, by making decisions and taking actions, may (or may not) save the world.

References

1. Pinker S (2018) Enlightenment now: The case for reason, science, humanism, and progress (1). Viking
2. Ord T (2020) The precipice: Existential risk and the future of humanity, Illustrated edition. Hachette Books
3. Harper TA (2024) Extinction panics is back, right on schedule. NewYork Times, January 28
4. Scheffer M, Carpenter SR, Lenton TM, Bascompte J, Brock W, Dakos V, van deKoppel J, van de Leemput IA, Levin SA, vanNes EH, Pascual M, Vandermeer J (2012) Anticipating critical transitions. Science 338(6105):344–348. https://doi.org/10.1126/science.1225244
5. Gladwell M (2002) He tipping point: How little things can make a big difference. Back Bay Books
6. Ferguson N (2018) The square and the tower, 1st edn. Penguin Press

Index

© The Editor(s) (if applicable) and The Author(s), under exclusive license to Springer
Nature Switzerland AG 2024
P. Érdi, *Feedback*,
https://doi.org/10.1007/978-3-031-62439-1